Schriftenreihe der Institute für Systemdynamik (ISD) und optische Systeme (IOS)

Editors-in-Chief
Jürgen Freudenberger, Konstanz, Deutschland
Johannes Reuter, Konstanz, Deutschland
Matthias Franz, Konstanz, Deutschland
Georg Umlauf, Konstanz, Deutschland

Die „Schriftenreihe der Institute für Systemdynamik (ISD) und optische Systeme (IOS)" deckt ein breites Themenspektrum ab: von angewandter Informatik bis zu Ingenieurswissenschaften. Die Institute für Systemdynamik und optische Systeme bilden gemeinsam einen Forschungsschwerpunkt der Hochschule Konstanz. Die Forschungsprogramme der beiden Institute umfassen informations- und regelungstechnische Fragestellungen sowie kognitive und bildgebende Systeme. Das Bindeglied ist dabei der Systemgedanke mit systemtechnischer Herangehensweise und damit verbunden die Suche nach Methoden zur Lösung interdisziplinärer, komplexer Probleme. In der Schriftenreihe werden Forschungsergebnisse in Form von Dissertationen veröffentlicht.

The "Series of the institutes of System Dynamics (ISD) and Optical Systems (IOS)" covers a broad range of topics: from applied computer science to engineering. The institutes of System Dynamics and Optical Systems form a research focus of the HTWG Konstanz. The research programs of both institutes cover problems in information technology and control engineering as well as cognitive and imaging systems. The connective link is the system concept and the systems engineering approach, i.e. the search for methods and solutions of interdisciplinary, complex problems. The series publishes research results in the form of dissertations.

Weitere Bände in der Reihe http://www.springer.com/series/16265

Mohammed Rajab

Channel and Source Coding for Non-Volatile Flash Memories

Mohammed Rajab
Konstanz, Germany

Dissertation Universität Ulm, 26. Juli 2019

ISSN 2661-8087   ISSN 2661-8095   (electronic)
Schriftenreihe der Institute für Systemdynamik (ISD) und optische Systeme (IOS)
ISBN 978-3-658-28981-2   ISBN 978-3-658-28982-9   (eBook)
https://doi.org/10.1007/978-3-658-28982-9

This Springer Vieweg imprint is published by the registered company Springer Fachmedien Wiesbaden GmbH part of Springer Nature.
The registered company address is: Abraham-Lincoln-Str. 46, 65189 Wiesbaden, Germany

For my parents, especially my mother soul.

Acknowledgements

Firstly, I would like to express my deepest gratitude to my supervisor Prof. Dr.-Ing. Jürgen Freudenberger for having faith in me, giving me the opportunity to work in his research group, and for supporting me in numerous ways. He has constantly provided me with inspiring scientific environment at the Institute for System Dynamics (ISD) of the University of Applied Sciences (HTWG) Konstanz, Germany. He has guided me through the development of this dissertation with his valuable suggestions and constructive discussions and continuous support throughout my thesis-writing period, furthermore, he encouraged me to enhance my writing skills. It was really my privilege and honour to gain from his exceptional scientific knowledge as well as his extraordinary human qualities. I would like also to thank Prof. Dr.-Ing. Antonia Wachter-Zeh for serving as the second reviewer and for her valuable comments.

Moreover, I would like to extend my special gratitude to Prof. Dr. Sergo Shavgulidze for his interesting ideas during his stay in Germany. In addition, I would like to show my deepest appreciation to all my colleagues at the ISD institute for their friendship, support and motivation. Moreover, I would like thank all my friends in Malaysia and Germany.

Many special thanks to Hyperstone GmbH, Konstanz and the German Federal Ministry of Research and Education (BMBF) for supporting this project and especially to BMBF for its financial support.

For those who have influenced me most, my family members, first and foremost I would like to express my sincere appreciation to my parents for their wise counsel, love, guidance, and encouragement throughout my life and study. I also wish to thank my brothers, my sisters, and my loving wife who encouraged me and prayed for me throughout the time of my work. Special thanks to my cousins who always stand with me. I am also grateful to my friends and neighbours in my hometown for their support.

Finally, I would like to dedicate this work to my late mother Mrs. Nima Rajab soul whose dreams and words have resulted in this achievement. You were a guarding angel calming me in hard times and feeling your prayers and love surrounding me in this hard world.

Contents

Abstract

NAND flash memory is widely used for data storage due to low power consumption, high throughput, short random access latency, and high density. The storage density of the NAND flash memory devices increases from one generation to the next, albeit at the expense of storage reliability.

Our objective in this dissertation is to improve the reliability of the NAND flash memory with a low hard implementation cost. We investigate the error characteristic, i.e. the various noises of the NAND flash memory. Based on the error behavior at different life-aging stages, we develop offset calibration techniques that minimize the bit error rate (BER).

Furthermore, we introduce data compression to reduce the write amplification effect and support the error correction codes (ECC) unit. In the first scenario, the numerical results show that the data compression can reduce the wear-out by minimizing the amount of data that is written to the flash. In the ECC scenario, the compression gain is used to improve the ECC capability. Based on the first scenario, the write amplification effect can be halved for the considered target flash and data model. By combining the ECC and data compression, the NAND flash memory lifetime improves three fold compared with uncompressed data for the same data model.

In order to improve the data reliability of the NAND flash memory, we investigate different ECC schemes based on concatenated codes like product codes, half-product codes, and generalized concatenated codes (GCC). We propose a construction for high-rate GCC for hard-input decoding. ECC based on soft-input decoding can significantly improve the reliability of NAND flash memories. Therefore, we propose a low-complexity soft-input decoding algorithm for high-rate GCC.

List of Acronyms

AWGN	additiv white gaussian noise
BCH	Bose Chaudhuri Hocqenhem
BER	bit error rates
BL	bit line
BMA	Berlekamp Massey algorithm
BSC	binary symmetric channel
BWT	Burrows-Wheeler transform
BMH	BWT-MTF-Huffman
BAC	binary asymmetric channel
CSB	center significant bit
CG	control gate
CCI	cell-to-cell interference
DMC	discrete memoryless channel
ECC	error correction codes
EOL	end of lifetime
EMG	exponentially-modified Gaussian
FNT	Fowler-Nordheim tunneling
FPGA	field programmable gate arrays
FG	floating gate
GCC	generalized concatenated codes
GEL	generalized error-locating
GF	Galois fields
GND	ground
GSL	ground selector line
GEL	generalized error-locating
LDPC	low-density parity-check
LLR	log-likelihood ratio
LZW	Lempel-Ziv-Welch
LUT	look-up tables

LSB	least significant bit
UB	upper bound
LDGM	low-density generator matrix
ML	maximum likelihood
MLC	multi-level cell
MTF	move-to-front
MH	MTF-Huffman
MOS	metal oxide silicon
KL	Kullback-Leibler
KB	Kilobyte
MSB	most significant bit
PDLZW	parallel dictionary LZW
P/E	program/erase
PC	product codes
HPC	half-product codes
RS	Reed-Solomon
RD	Read disturb
RAM	random access memory
SLC	single-level cell
SNR	signal to noise ratio
SPC	single parity-check
SSD	solid-state drive
SSL	sensing selector line
TLC	triple-level cell
UP	upper page
WL	word line
WA	write amplification
WER	word error rates

1 Introduction

Nowadays, flash memory and especially NAND flash memory are very important storage medium in industrial application. NAND flash memory provides high-density, low-latency, fast-programming and erase operation speeds at low costs. Flash memories are mechanical shock-resistant non-volatile memories. Therefore, flash memories are very interesting for many devices that require high data reliability, e.g. for industrial robotics, scientific and medical instrumentation.

During the last decade, the storage density has rapidly increased. However, higher density leads to degradation in the flash memory reliability and lifetime. The information may be read erroneously, where the error probability depends on the storage density and noises caused by several effects, including program/erase (P/E) cycle effects, data retention, program/read disturb, cross temperature and cell-to-cell interference (CCI).

1.1 Goals

The main aim of this dissertation is to improve the data reliability of NAND flash memories. This goal can be achieved in several ways, e.g. using data compression and error correction codes (ECC). In particular, we take industrial requirements into account. Such industrial applications of flash memories require very low residual error rates, high code rates, very high throughput, low latency, and efficient algorithms with low complexity. We propose solutions based on the flash memory behavior. In order to understand the flash memory channel, we investigate several flash memory types. The research goals of this dissertation are listed as follows:

- Several physical effects shift the threshold voltage distribution during the flash operations, which increases the bit error rates (BER). We investigate offset calibration techniques, where the aim is to minimize the BER with low decoding latency and complexity.

- Flash memory can wear-out after a certain number of program and erase operations. In order to erase a block, the pages that are remapped to a new block require background writes. This leads to the write amplification (WA), which reduces the flash memory reliability. In order to reduce the WA, data compression schemes can be used to reduce the amount of user data. These schemes shall exhibit industrial requirements such as efficient performance and low complexity.

- Increasing the storage density makes ECC an important component in flash controllers. Therefore, we investigate several concatenated schemes that can satisfy industrial requirements like half-product codes (HPC) and generalized concatenated codes (GCC). The objective of using concatenated schemes is to achieve low residual error rates. ECC based on GCC has strong potential for various applications in data communication and data storage systems, e.g. for non-volatile flash memories. The residual error rates for GCCs can be determined analytically, which is important for applications where a low probability of failure has to be guaranteed. The aim is to construct high-rate codes with low complexity that can be applicable for flash memory.

- ECCs are widely used in flash memories to ensure data integrity and reliability. The choice of an ECC scheme can be determined by the industrial requirements. A strong ECC scheme requires a larger amount for data redundancy. The main aim is to increase the error correction capability by using data compression. This can be achieved by combining the data compression with ECC, which should meet industrial requirements.

- Soft-input decoding can clearly improve the decoding performance. New flash types require low-complexity soft-input decoding methods. Furthermore, soft-input decoding requires reliability information about the state of the memory cell. This can be achieved by using the pilot data, which requires many pilot bits to estimate the corresponding log-likelihood ratio (LLR) values for lower error probabilities. The objective is to avoid such additional pilot data for channel estimation.

1.2 Structure of the dissertation

This dissertation is organized as follows.

In *chapter 2*, we provide a basic background about the flash memory functionality and we investigate different error characteristics during different flash aging states. Based on flash measurements, we propose an offset calibration mechanism with low latency and complexity.

In *chapter 3*, we propose lossless data compression algorithms that are applicable for flash memory. These universal schemes reduce the WA effect for particular flash memory and data models.

In *chapter 4*, we demonstrate a coding scheme that combines the data compression with ECC. This scheme improves the lifetime of the NAND flash and can be implemented with moderate hardware requirements.

In *chapter 5*, we introduce the preliminary for several concatenated codes based on low-complexity extended Bose Chaudhuri Hocqenhem (BCH) codes. For lower error-correcting BCH codes, we propose an inversion-less version of Peterson's algorithm and its decoding architecture. Later, we present the GCC encoding, decoding scheme as its code parameter design and error bound. We propose new construction where single parity-check (SPC) codes are used in the first level of the GCC. This enables high-rate codes and low complexity.

In *chapter 6*, we propose a soft-input decoder for high-rate GCC. Therefore, we demonstrate a soft-input decoding for the inner BCH codes based on a bit-flipping decoder. This method uses a fixed number of test patterns and an algebraic decoder for soft-decoding. Moreover, we propose an acceptance criterion for the final candidate codeword. This acceptance criterion improves the decoding performance and reduces the decoding complexity.

Finally, a conclusion to the dissertation is provided in *chapter 7*.

2 Error characteristics and read threshold calibration for flash memories

Nowadays, flash memory is commonly used in persistent storage devices. Increasing storage density and faster reading and programing speed lead to a degradation in flash memory reliability and lifetime. In this chapter, we introduce the basic background to NAND flash memory, which guides understanding the flash memory reliability problem.

First, we discuss how NAND flash memory is organized and its operations in section 2.1. Next, we provide an analysis of the error characteristic in triple-level cell (TLC) flash memory based on measurements in section 2.2. Moreover, we investigate the error characteristic for the flash memory at different lifetimes. These errors are caused by several effects that degrade the performance of the flash memory; for example, P/E cycles and data retention. Due to these effects, the original threshold voltage is shifted. With pre-defined fixed read thresholds, a voltage shift increases the BER.

Most modern flash memories use a mechanism called read-retry to adjust the read threshold voltages at the minimum BER, which is used to read data from a cell. In section 2.4, we propose offset calibration mechanisms to compensate voltage shifts. This mechanism aims to reduce the BER and the reading latency by using the meta-data protected by the ECC in the flash controller. The offset calibration mechanism aims to find the near-optimal read threshold voltage based on a few hundred bits of the meta-data. Finally, we summarize this chapter in section 2.5. Parts of this chapter are published in [1, 2, 3, 4]

2.1 NAND flash memory basics

Flash memories are becoming increasingly important for mass data storage, due to the increasing number of devices with large storage requirements. Moreover, they are non-volatile, i.e. they can retain information even without a power supply. Flash memory was first introduced in NOR devices, and later NAND flash memories were introduced by Toshiba in 1984. The major difference between NOR and NAND flash memory is that the cells are connected in parallel for NOR flash, whereas in NAND flash cells are connected in series. Today, NAND flash memories become important components in embedded systems as well as consumer electronics. Flash memory is considered in many applications ranging from enterprise storage (solid-state drive (SSD)) to personal devices, due to high access speed and low power consumption.

The uncomplicated structure of NAND flash memory enables low cost and high capacity. NAND flash memories outperform traditional magnetic hard disks due to their lower access latency. Furthermore, NAND flash memories can be adapted to industrial applications that require fast and reliable storage mechanisms. The recent transition from two-dimensional (2D) to three-dimensional (3D) flash memories provides higher density. However, the 3D flash array has a more complex architecture compared with the 2D flash arrays [5]. This development in the flash memories has a major impact on the reliability.

2.1.1 NAND flash memory architecture

In flash memories, the basic memory element is a floating gate (FG) transistor. Figure 2.1 presents the FG cell, which is similar to a metal oxide silicon (MOS) field effect transistor or MOSFET. The FG is electrically isolated where the charge remains trapped inside it. The FG states can be changed by Fowler-Nordheim tunneling (FNT) [6]. The FG cell has three layers: on the top of the cell is the control gate (CG), the oxide layer which holds the floating gate from both sides and the substrate layer at the bottom. The NAND flash memory is arranged as $2D$ or $3D$ array of flash cells, where the block is the smallest unit that can be erased, and the page is the smallest unit for read and program operations. Typical block sizes are 512 to 4 Kilobyte (KB) with additional meta information.

Figure 2.1 presents the cell arrangement of NAND flash memory, where the vertical strings of cells are connected in series by sharing the same bit line (BL). In the other dimension, all gates are connected in parallel to a word line (WL), which represents one page. A flash block comprises all pages that are connected with the same BLs. The ground selector line (GSL) and the sensing

selector line (SSL) are important to apply the different operation modes, which
are explained in the next sub-section.

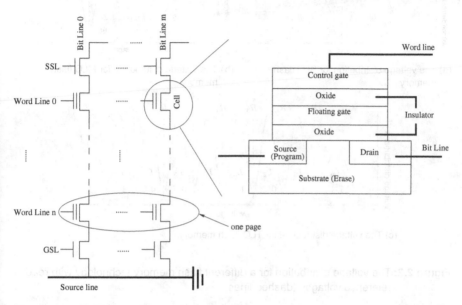

Figure 2.1: NAND flash memory cells arrangement and the structure of the cell.

Flash devices are organized by the number of bits that can be stored in the cell.
Each cell contains one FG transistor. which can store at least one bit in the
form of a charge level. This charge determines the threshold voltage V_{th} of the
flash cell. Figure 2.2 depicts exemplary voltage distributions for different flash
memory technologies. In this figure, the x-axis represents threshold voltages
V_{th} and the y-axis represents the probability distribution of programmed voltages that correspond with different charge states.

Here, the threshold voltage distribution of the FG is modeled as a Gaussian
distribution. The single-level cell (SLC) devices store one bit per cell, where L_0
is the erased state and L_1 is the programmed state. Furthermore, figure 2.2(b)
depicts multi-level cell (MLC) devices that can store two bits per cell, where L_0
is the erased state and (L_1, L_2, and L_3) are the programmed states. Each
level ($L_0, \ldots, L_3$) encodes a 2-bit value that is stored in the flash cell (e.g. 11,

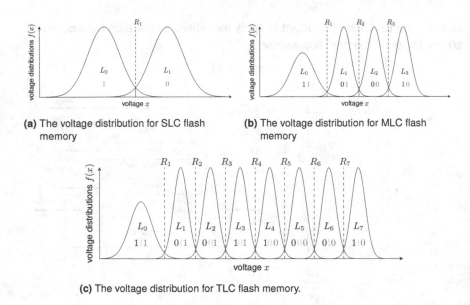

(a) The voltage distribution for SLC flash memory

(b) The voltage distribution for MLC flash memory

(c) The voltage distribution for TLC flash memory.

Figure 2.2: The voltage distribution for a different flash memory technology with read reference voltages (dashed lines)

01, 00, and 10). The first bit is the most significant bit (MSB) marked in blue color and the second bit is the least significant bit (LSB), marked in red color.

TLC refers to cells that can store three bits. Each level ($L_0, \ldots, L_7$) encodes a 3-bit value that is stored in the flash cell (e.g. 111, 011, 001, 101, 100, 000, 010, and 110). The first bit is the MSB with blue color, the second bit is the center significant bit (CSB) with green color, and the last bit is the LSB with red color. Typically, the LSB, CSB, and MSB are mapped to different pages. Note that some manufacturers may use different mapping values for different levels; for example, (111, 110, 100, 000, 010, 011, 001, and 101).

2.1.2 NAND flash memory operations

The flash memory cell has three main operations,namely erase, program, and read. figure 2.3 presents different NAND flash memory operations. We briefly explain the three basic operations.

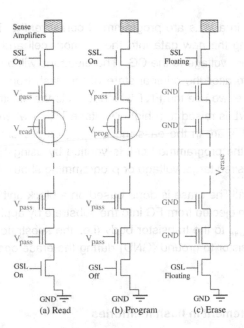

Figure 2.3: Voltages applied to flash cell to perform (a) read, (b) program, and (c) erase operations.

- Read operation: The read is done page-by-page. In figure 2.3(a), the read voltage (V_{read}) is applied to the selected WL while the other WLs are fixed to V_{pass}. To read the data stored in an SLC, we need a single reference threshold voltage R_1 to distinguish the corresponding bit value of "1" and "0" as shown in figure 2.2(a). Likewise, to read the data stored in MLC, it is required to have three thresholds to distinguish between ($L_0, \ldots, L_3$). In figure 2.2(b), read reference voltage R_2 determines the LSB, whereas two read reference voltages (R_1 and R_3) are required for the MSB. figure 2.2(c) presents seven reference voltages ($R_1, \ldots, R_7$) for TLC. With the mapping in figure 2.2(c), only one read reference voltage needs to be applied to the cell to read the LSB, whereas two read reference voltages are necessary for the CSB. For the MSB, four read reference voltages need to be applied in sequence.

- Program operation: The programming is conducted based on a page unit, where all cells in a WLs are programmed concurrently. The program operation is writing the new data into the memory cell and is performed by applying a higher voltage to the CG. The electrons will be injected into the FG. In order to alter the original state of the flash from the erased state to program state, we use the FNT which tunnels the electrons into the FG. The selected WL is raised to a higher-voltage V_{prog}, where the unselected gates on the BLs are in the pass mode V_{pass} as present in figure 2.3(b). Subsequently, the programmed cell is verified by using V_{verify} whether it reached the desired target voltage or programming should be repeated.

• Erase operation: The erase is done based on a block unit. Using FNT, the electrons will be ejected from FG into the substrate by applying a high erasing voltage V_{erase} to the transistor body (i.e. the substrate) and setting the CG of the transistor to ground (GND) during the erase operation as shown in figure 2.3(c).

2.2 Error measurement in flash memories

In this section, we discuss the reliability degradation of the NAND flash memory during the operations of the flash. The reliability of the flash memory suffers from various error causes like P/E cycles, data retention, program errors, errors due the number of reads after programming (read disturb) [7] and CCI. These effects cause the threshold voltage distribution to widen. Furthermore, the read reference threshold is changed. Hence, the variance varies from state to state, where some states are less reliable. This results in different error probabilities for the LSB, CSB and MSB pages. Moreover, the error probability is not equal for zeros and ones, where the error probabilities can differ by more than two orders of magnitude [8]. Ultimately, all of these effects may increase the BER [1].

We demonstrate measurements based on TLC NAND flash memory devices, which are denoted as vendor-A. We briefly discuss each type of error independently based on this device.

2.2.1 Voltage distributions with P/E cycles

Flash memories have a limited number of P/E cycles. The high voltages required for P/E operation and a high electric field cause small damages to the cell during every cycle. Repeating P/E cycles will cause damage in the tunnel oxide due to electrons becoming trapped within it [9]. Consequently, the cell becomes difficult to erase.

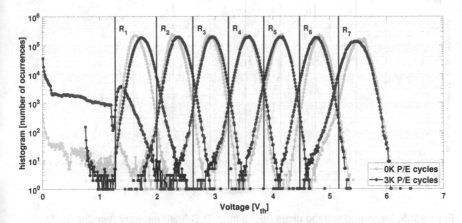

Figure 2.4: Measured voltage distributions for a TLC flash memory (vendor-A). The bold vertical lines represent the default read reference voltage after the first programming. The figure presents the change of the distribution due to cycling (after 3000 P/E cycles).

Figure 2.4 depicts measured voltage distributions for TLC flash memory (vendor-A) after 3000 P/E cycles compared with the distribution after the first programming (i.e. 0K P/E cycles, which is also called fresh block). The vertical lines present the default threshold voltages $(R_1, ..., R_7)$. Note that this measurement is based on the data of a complete block. As can be seen in figure 2.4, the distribution after 3000 P/E cycles has a shift to higher voltages for lower levels, whereas higher levels (i.e. L_6 and L_7) have a shift to lower voltages. For the erased level L_0, the distribution has a larger number of occurrences due to the charge trapping/detrapping in the oxide. Furthermore, the distribution of all levels have a large tail.

2.2.2 Voltage distributions with data retention

Data retention is the parameter that specifies the maximum period after programming that data can be expected to be retrieved from the memory cells [10]. In other words, retention errors happen due to charge leakage over time after a cell is programmed [11].

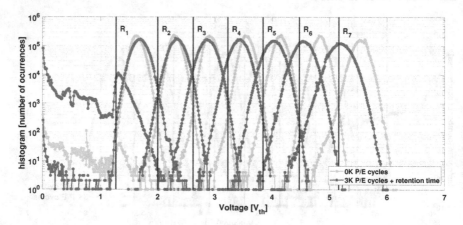

Figure 2.5: Measured voltage distributions for a TLC flash memory (vendor-A). The bold vertical lines represent the default read reference voltage after the first programming. The figure indicates the additional charge loss do to data retention (end of lifetime (EOL)).

In many applications, data retention may be more critical than flash endurance. Long data retention times cause charge losses and a change of the threshold voltage distribution. The data retention performance of MLC and TLC devices are considerably lower than SLC.

Figure 2.5 depicts measured voltage distributions for a TLC flash memory after 3000 P/E cycles and one year of data retention (data retention was simulated by a baking procedure [12]). In this case, the TLC flash memory reaches the EOL. The EOL is reached when the device reaches the maximum number of P/E cycles and a specific data retention performance. In figure 2.5, the charge loss is clearly visible in all charge levels. This charge loss induces a shift of the threshold voltage distributions to the lower voltages, (i.e. the threshold voltages should be shifted to the left which minimizes the BER).

Table 2.1: BER for TLC flash memory block (Vandor-A) at fresh block, 3K P/E cycles, and at the EOL.

page	fresh block	3K P/E cycles	EOL
LSB	4.8E-4	2.4E-3	1.6E-2
CSB	4.2E-4	7.7E-4	2.9E-2
MSB	1.1E-3	2.4E-3	5.0E-2

The P/E cycles have a major impact on the charge loss, where the charge loss is increased with more P/E cycles. Table 2.1 shows the BER at the 3K P/E cycles and at the EOL. We assume that the state L_j can only be falsely detected as state L_{j-1} or state L_{j+1}, i.e. $p(L_i|L_j) = 0 \ \forall \ |i - j| > 1$, where L_i is the detected state and L_j is the real state. Due the P/E cycles over time, the increasing of BER occurred in all pages with different voltage shifts.

A shift of the threshold voltage may result in higher error probabilities and an asymmetric error characteristic. This is indicated by the measurements in figure 2.6, where the bit error probability for the same TLC is plotted versus the shift of the voltage threshold. The three markers indicate the voltage shift for different cycling and data retention conditions. The value of $0V$ corresponds to the original read reference voltage after the first programming.

Figure 2.6 presents the BER for the highest L_7 and second highest L_6 charge level after the first programming. The two curves in figure 2.6 indicate the bit error probability for zeros and ones depending on the read reference voltage. A deviation from the optimal read reference voltage results in an increased error probability. It can be observed that the bit error probability is symmetric at the ideal reference voltage and becomes asymmetric with increasing voltage shifts. Similar to the approaches in [13, 14], we exploit this asymmetry of the error probabilities for the adaptation of the thresholds.

Based on the P/E cycles and data retention measurement of vendor-A, we can observe the following:

- the threshold voltage distributions of the higher states shift to lower voltages with retention time;

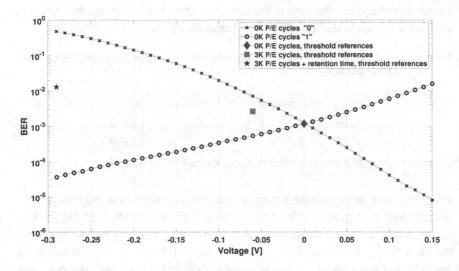

Figure 2.6: Bit error probability versus the threshold voltage shift.

- the threshold voltage distribution of each state becomes wider with higher retention time;
- the threshold voltage distribution of a higher-voltage state shifts faster than that of a lower-voltage state; and
- the higher threshold voltage distributions change more significantly over retention time compared with the lower threshold voltage distributions.

Offset calibration techniques may be used to compensate the shift of the voltage threshold [15]. However, threshold calibration is not available for all flash devices. As indicated in [8] and by the presented measurements, the error characteristic of the flash memory is asymmetric without calibration.

2.2.3 Read/Program disturb and cross temperature

Read disturb (RD) is a phenomenon caused by reading neighboring cells [7, 16]. TLC flash devices are more prone to read/program disturb compared with SLC and MLC. In order to read data from a cell, a high pass-threshold voltage

V_{pass} has to applied to switch on all other cells in the same bit line, where only one page can be read at the time. Accordingly, the threshold voltage of the neighboring pages might be shifted to a higher voltage.

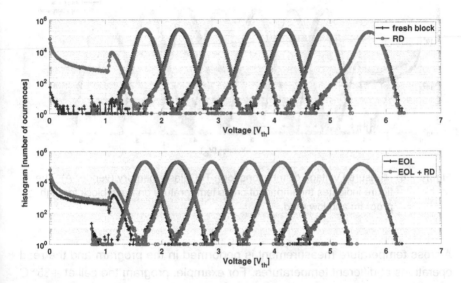

Figure 2.7: Measured voltage distributions for a TLC flash memory (Vandor-A). The upper figure indicates the impact of RD on fresh block and the lower indicates the impact of the RD at the EOL.

Figure 2.7 presents the measurement of the RD for a fresh block and under P/E cycles effect. The upper figure illustrates the voltage distribution of a fresh block and after RD (i.e. after 10k read cycles). The lower figure presents the voltage distribution at the EOL and after the RD of 3k read cycles. In both cases, we observed that the RD affects the lower charge levels by shifting the read threshold to the higher voltages, especially (L_0) and slightly $(L_1$ and $L_2)$. In both cases, the BER is increased for all pages, even without any change in the default threshold of the higher charge levels.

During the program process on the cell, a high program voltage is applied to the chosen WL, where the neighboring cells in the same WL might cause a shift due to parasitic coupling. Similarly, the flash memory reliability may suffer from temperature-related charge loss [7].

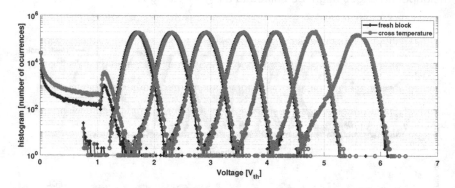

Figure 2.8: Measured voltage distributions for a TLC flash memory (vendor-A). The figure indicates the impact of cross temperature on fresh block for high program and low read.

A cross temperature measurement is performed in the program and the read operations at different temperatures. For example, program the cell at $+85\,^{\circ}C$ and read the cell at $-40\,^{\circ}C$ or the opposite condition. Figure 2.8 illiterates voltage distributions for a fresh block and under cross temperature effect, (i.e. program at $+85\,^{\circ}C$ and read at $-40\,^{\circ}C$). This in turn will cause a shift of the threshold to higher voltages, as observed from figure 2.8. In terms of BER, program at $+85\,^{\circ}C$ and read at $-40\,^{\circ}C$ seems more critical than the opposite condition.

2.2.4 Cell-to-cell interference

CCI is a phenomenon between adjacent cells. After applying the programming voltage to the cell, the injected electrons will be coupled with the programmed voltages of the adjacent cells due to parasitic capacitors [17].

Figure 2.9 illustrates the CCI, where the adjacent cells influence the victim cell. There are two different types of program interference which can be WL-to-WL program interference and BL-to-BL program interference. Many researchers

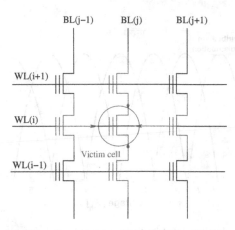

Figure 2.9: The adjacent cells can induce program interference on a victim cell.

have studied CCI during the last decade in an attempt to avoid or reduce the impact of the neighboring cells. As one example, using (All-Bit-Line) architecture enables reducing the CCI of the victim cell due to page-order programming, where different bits in the cell are programmed separately [18, 19]. This phenomenon is not considered in our research.

Ultimately, the reliability of the NAND flash memory is affected by other factors over lifetime, such as radiation and media defect.

2.3 Fitting the threshold distribution to the theoretical models

Based on the error characteristic of TLC flash, we can utilize the channel model theoretically. Here, we present the Gaussian distribution, exponential distribution, and exponentially-modified Gaussian (EMG) distribution, all of which can describe the error characteristic of the flash. Practically, we have two TLC examples of the same flash memory, i.e. vendor-A, but different samples and error characteristics at the EOL.

Figure 2.10 depicts a measured voltage distribution of a chip from vendor-A at the EOL. Based on the measurement of TLC flash, the probability density

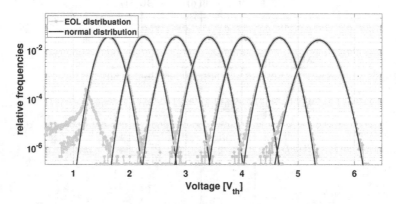

Figure 2.10: Measured voltage distributions for a TLC flash memory compared with the normal distribution.

function $f(x)$ of the threshold voltages is modeled by a Gaussian distribution with variance σ^2, mean μ, and probability density [20, 21].

$$f(x) = \frac{1}{\sqrt{2\pi\sigma^2}} e^{\frac{-(x-\mu)^2}{2\sigma^2}}. \tag{2.1}$$

Hence, the channel model for flash cells is equivalent to an additiv white gaussian noise (AWGN) channel where the noise variance σ^2 depends on the stored bit, i.e. the mean μ of the charge level.

The AWGN channel can be modeled for some flash type but perhaps not for others. Even some samples from the same flash type can be considered as AWGN distribution and others as exponential distribution due to single-sided exponential tails.

Figure 2.11 shows that the exponential distributions in the error regions are dominant compared with the normal distribution. Based on the measurement data from vendor-A at EOL, the distributions feature significant single-sided exponential tails, which are nearby the intersection point of the two adjacent distributions. Note that the first state (erase state) is not presented due to unreliable information. Figure 2.11 exhibits that the wear-out can widen the distribution with exponential tails especially on the left side of the distribution [22, 23, 24].

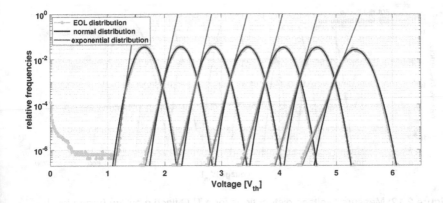

Figure 2.11: Measured voltage distribution for TLC flash memory of vendor-A (green level) compared with the normal distribution and the exponential distribution.

This exponential tail can be explained by charge trap in the cells especially at the EOL. Note that some voltage distributions can have physical effect at the early life.

Theoretically, we calculated the voltage distribution $g(x)$ as the exponential distribution,

$$g(x) = \begin{cases} \lambda \, e^{\lambda(x-x_0)} & , \ x \leq x_0 \\ 0 & , \ x > x_0 \end{cases} \tag{2.2}$$

where λ is the channel parameter, which functions as a metric for the charge trap density. Note that the exponential distribution is not held to infinity, but rather is limited by x_0.

Furthermore, the EMG distribution describes the sum of independent normal and exponential random variables. For instance, the probability density function of the EMG distribution that has a characteristic positive skew is

$$z(x) = \frac{\lambda}{2} \, e^{\frac{\lambda}{2}(2\mu+\lambda\sigma^2-2x)} \, \text{erfc}(\frac{\mu+\lambda\sigma^2-x}{\sqrt{2}\sigma}), \tag{2.3}$$

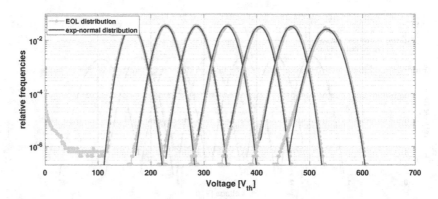

Figure 2.12: Measured voltage distributions for a TLC flash memory (green level) compared with the exponentially-modified Gaussian distribution.

where the parameters σ, μ, and λ evolve over both time and P/E cycling process, and the erfc is the complementary error function, defined as

$$\mathrm{erfc}(x) = \frac{2}{\sqrt{\pi}} \int_x^\infty e^{-t^2} dt. \tag{2.4}$$

Figure 2.12 demonstrates that the convolution of the normal and exponential probability density functions holds for vendor-A flash. The figure shows that this distribution fits the most voltage distributions at the EOL, where the distributions widen due to the wear-out. This implies that the σ increases compared with early life i.e. 0k P/E cycles as observed from the measurement distribution. Moreover, the distributions shift to the lower voltage due to the charge loss where the μ value decreases compared with 0k P/E cycles. Note that the μ value more strongly decreases for higher threshold voltage distributions compared with the lower threshold voltage distributions.

2.4 Read threshold calibration for flash memories

Flash memories may apply a read-retry mechanism to dynamically adjust the threshold voltage. First, the flash controller reads the data with the default read reference $V_{default}$. The flash controller sends the data to the ECC decoder. If the ECC decoder corrects the errors, there is no need to change the default value $V_{default}$. Otherwise, the flash controller will read several times using a certain set of reading threshold voltages until the ECC decoder can correct the errors, or fails to read the data.

Several papers have explored dynamic threshold adaptation concepts [13, 14, 25, 26], which adjust the read thresholds to minimize hard decoding BER. These dynamic threshold adaptation schemes can significantly reduce the BER compared with pre-defined fixed read thresholds. However, these schemes are based on assumptions that may not be satisfied for today's flash memories. In [13, 14], coding schemes are proposed that exploit the asymmetric error probabilities of the channel, e.g. based on Berger codes. Berger codes are able to detect any number of unidirectional errors. In the case of a large charge loss, it is possible to perform several reads with decreasing thresholds until the code detects a sufficiently low number of errors. This technique requires either long codes for the threshold calibration or is only applicable for very high error rates. In [26], a parameter estimation approach is proposed and an optimal threshold adaptation policy derived. This policy is based on the assumption of Gaussian voltage distributions. This assumption may not hold for all flash types, e.g. the cell threshold voltage distributions of MLC and TLC are actually highly asymmetric with exponential tails [27].

In order to optimize a page-wise read operation, we propose a near-optimal method based on the read-retry approach. The aim of the threshold calibration is to minimize the bit error probability. The bit error probability can be estimated directly based on known data, e.g. pilot data that is stored in the flash memory. Such pilot data may be available in the form of meta-data that is stored along with the user data. The meta-data is typically protected by strong ECC to ensure that this information can be read even in the worst-case situations. However, the error probabilities of TLC flash memories are typically in the range of 10^{-4} to 10^{-2}. Hence, large pilot data sets would be needed to estimate low BER values, whereas typically only a few hundred bits of meta-data are available per page. In this section, we propose the read-retry mechanism based on meta-data of the ECC.

2.4.1 Read-retry based on meta-data

The read-retry mechanism can reduce the number of errors by adjusting the threshold reference at the minimum BER of the two neighboring voltage distributions. Figure 2.13 presents the sum of the two neighboring voltage distributions for the MSB page where the MSB contains two threshold references, i.e. R_3 and R_7. In figure 2.13, the upper figure corresponds to R_3 and the lower figure corresponds to R_7. This MSB has been measured for different conditions: at 1.5k P/E cycles and 13 hours of data retention time, at 1.5k P/E cycles and 80 hours of data retention time, and at the EOL, respectively.

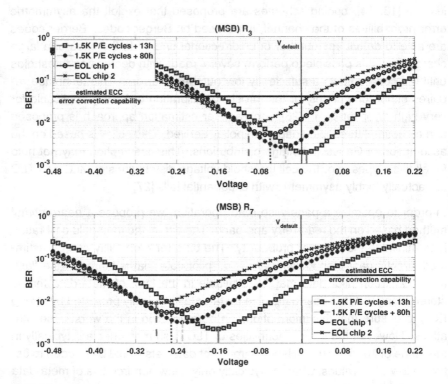

Figure 2.13: Measured BER of MSB for different lifetime of a TLC flash memory.

The vertical line at $0V$ presents the default threshold $V_{default}$, where the other vertical lines represent the optimal threshold voltage V_{opt} for each condition of R_3 and R_7, respectively. The minimum value of the voltage distribution represents the area that has the minimum number of errors, which is the crossing point of the two neighboring voltage distributions. The horizontal bold line represents the approximation of ECC correction capability, i.e. the upper bound. At 1500 P/E cycles and 13 hours retention time, R_3 does not need any read-retry mechanism or offset calibration techniques. R_7 has a threshold shift of about $-0.16V$, but the error probability still does not exceed the ECC correction capability. The $V_{default}$ is not optimal for other cases with higher data retention. For instance, R_7 requires adjusting the threshold reference because the error probability larger than ECC correction capability.

The read-retry method aims to reduce the number of read-retry iterations, i.e. the reading latency. The flash controller searches for the optimal threshold voltage V_{opt} by counting the number of errors in the meta-data for each read operation. Note that if the number of errors exceeds the ECC correction capability, the flash controller is unable to count how many errors exist in the meta-data ECC.

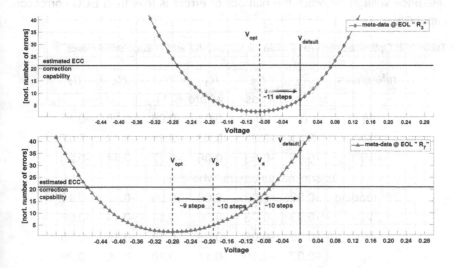

Figure 2.14: Measured voltage distributions of MSB for raw data

Figure 2.14 presents the meta-data distribution of MSB at EOL, where this distribution is based on the average of 256 pages. The y-axis presents the normalized errors, where the ECC unit can correct up to error correction capability t. In this figure, R_3 does not need the optimal threshold because the number of errors in the meta-data is less than t. However, the default threshold is not optimal for R_3, where we need to shift the $V_{default}$ by about $-0.09V$ to be at V_{opt}. In the case of R_7, the $V_{default}$ is not optimal because the number of error exceeds t. In order to find the optimal threshold, perhaps more than one read is required. Figure 2.14 shows that R_7 requires three read-retry iterations (i.e. V_a, V_b, and V_{opt}) to find the optimal threshold. In this case, the ECC decoder fails to decode the meta-data of the MSB at the $V_{default}$ because the sum of error for R_3 and R_7 exceeds t. It is very important to find the optimal threshold, where the reliability of TLC flash can be improved. In order to improve the read-retry mechanism, it requires the following:

- an optimized read-retry voltage shift table;

- strategies based on the error conditions acceptance.

These requirements aim to minimize the read operation latency and find a read reference voltage for which the number of errors is less than ECC correction capability.

Table 2.2: Optimized read-retry table for successful and unsuccessful cases

references	R_2	R_3	R_4	R_5	R_6	R_7
successful cases, where $E(1) \leq t$						
V_1	0	0	-0.01	-0.08	-0.05	-0.14
V_2	-0.01	-0.01	-0.03	-0.14	-0.10	-0.17
V_3	-0.05	-0.03	-0.08	-0.17	-0.14	-0.23
unsuccessful cases, where $E(1) > t$						
2nd reading	-0.02	-0.03	-0.05	-0.09	-0.10	-0.10
V_4	-0.03	-0.03	-0.08	-0.14	-0.14	-0.23
V_5	-0.05	-0.06	-0.09	-0.18	-0.16	-0.25
V_6	-0.07	-0.10	-0.14	-0.20	-0.19	-0.28

As we notice from figure 2.13, the ECC decode the data using the default threshold $V_{default}$ successfully in some cases, although the number of errors

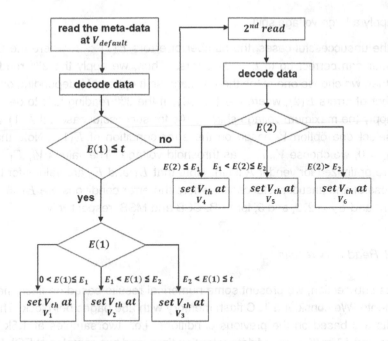

Figure 2.15: Flowchart of read-retry strategy for all pages

remains high. In order to optimize successful and unsuccessful cases, we propose a strategy based on two reading operations. Figure 2.15 illustrates the flowchart of the strategy, where $E(k)$ is the number of errors for k reading. This strategy chooses one of multiple voltage shift options V_i based on the number of errors in the meta-data. Table 2.2 shows the voltage shifts V_i, where $i \in \{1, \dots, 6\}$ for each read-retry. The conditions of the voltage shifts are shown in figure 2.15. R_1 is always chosen at $V_{default}$. Note that the pages require multiple different voltage shifts for their mapped references R_j at the same time.

The main idea of this strategy is to distinguish between a large and small shift by applying the 2nd reading with a small shift for unsuccessful cases. This small shift will provide information about the flash status, and it guides tracking the optimal threshold. For instance, successful cases only require setting the threshold voltage at V_i based on $E(1)$ of the 1st reading, whereas unsuccessful cases are based on $E(2)$ of the 2nd reading. Note that if the 2nd reading fails, we apply a large voltage shift.

For the unsuccessful cases, the number of errors $E(1) > t$, where the ECC decoder can correct up to $t = 21$ errors. Thus, we apply the 2nd reading, and then we choose one of the three voltage shift options V_i depending on the number of errors $E(2)$, where $i \in \{4, 5, 6\}$. If the 2nd reading fails to decode, we apply the maximum voltage shift V_6. As for successful cases, if $E(1) \neq 0$, we select one option V_i based on the error condition of $E(1)$. Note that if $E(1) = 0$, we choose $V_{default}$ as threshold voltage. The values V_i, E_1, and E_2 are optimized for vendor-A flash. Note that E_1 and E_2 are values for both successful and unsuccessful cases, where the error conditions are $E_1 = 8, 6$, and 13 and $E_2 = 2, 2$, and 5, for LSB, CSB and MSB, respectively.

2.4.2 Read-retry results

In this sub-section, we present some numerical results based on flash measurements. We consider a TLC flash memory with 256 pages per block. These results are based on the previous conditions, i.e. two samples at 1.5k P/E cycles and 13 or 80 hours of data retention time, and two samples at EOL from different chips, respectively. As we observe from the data measurements, we can use a fixed threshold value at the early life even in the higher references (i.e. based on the optimal threshold of the first page).

For EOL cases, the read-retry is required for all pages, especially for the higher references. Note that the numerical results are calculated based on the conditional probability. For example, in figure 2.2(c), the BER of R_7 is an approximation of the conditional probability of making an error in one of L_6 or L_7, assuming that the other charge levels have negligible error probability. This can be considered as the worst-case scenario, where the charge levels with higher reliability are neglected, for instance L_5. In general, the conditional error probability of R_j is approximated by the error rate of the neighboring charge levels, i.e. L_{j-1} and L_j.

Figure 2.16 shows the BER results for all references R_j at the EOL for a fixed threshold, the proposed read-retry, and the optimal read threshold for each page. For R_1, we use a fixed read threshold for all pages. As can be seen, the BER degrades with the fixed threshold for R_4, R_5 and R_7, whereas R_2, R_3 and

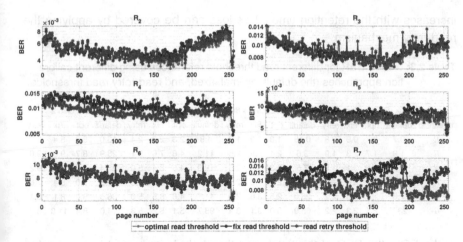

Figure 2.16: BER for all pages at EOL, chip 2 with the optimal, fixed threshold, and retry threshold.

R_6 have a good performance by using the fixed threshold. This implies that the higher references require a read-retry mechanism based on each page, whereas the lower references can be applied based on the optimal threshold of the first page.

The read-retry mechanism tracks the voltage shifts of all references and reduces the BER compared with the default read threshold $V_{default}$. The proposed schemes are based on 510 bits of the meta-data for each page. In the lower reference, the read-retry mechanism requires lager meta-data due to the low reliability. This can be achieved by summing up the meta-data from neighboring pages. Based on the flash manufacturing, we can determine the maximum number of pages that can be summed.

Table 2.3 presents the BER, which is averaged over all pages of the TLC flash memory at different conditions. It shows that the default read threshold $V_{default}$ is not suitable for the EOL cases and for the higher threshold voltages of 1.5k P/E cycles cases.

In the early life, i.e. 1.5k P/E cycles and 13 hours, the fixed read threshold has a small performance loss compered with the read-retry mechanism, where this loss is small compared with the decoding latency of each page. The BER increases with the retention time. This loss can be covered by applying the read-retry mechanism.

Table 2.3: $\times 10^{-3}$ BER for different read threshold references using different calibration approaches (the default, optimal, fixed, and read-retry read thresholds)

Samples	1.5k (P/E) cycles + 13h				1.5k (P/E) cycles + 80h				EOL, chip 1				EOL, chip 2			
threshold	default	opt.	Fix	retry	default	opt.	Fix	retry	default	opt.	Fix	retry	default	opt.	Fix	retry
R_2	0.7	0.7	0.9	0.8	0.16	1.0	1.3	1.2	4.8	2.3	2.9	2.4	13.0	5.0	5.4	5.2
R_3	1.9	1.7	2.3	1.9	3.6	2.4	2.5	2.6	11.2	4.6	7.6	5.1	28.9	8.7	9.0	9.0
R_4	2.2	1.6	3.1	1.8	7.5	2.3	3.0	2.6	19.3	4.5	8.4	5.0	53.4	8.9	11.2	9.2
R_5	8.4	1.7	3.6	2.9	26.3	2.4	4.0	2.7	46.6	4.1	9.7	4.6	92.5	7.0	8.4	7.6
R_6	7.8	1.7	1.8	2.2	28.0	2.5	3.8	2.9	47.7	4.6	8.5	5.0	86.5	8.0	8.1	8.1
R_7	23.7	2.1	2.3	2.2	65.9	3.0	3.1	3.4	98.7	5.1	5.2	5.3	14.0	8.4	11.8	8.7

At the EOL, the BER of the fixed read threshold is increased in some references R_j compared with the read-retry mechanism for both chips. For instance, R_2, R_3, R_6 from chip 2, the fixed read threshold achieves good results, but not for other references. For R_4, R_5, and R_7, the read-retry mechanism clearly reduces the BER.

2.5 Summary

In this chapter, we have presented the basic architecture of the NAND flash memory. Over the lifetime, the reliability of the flash memory suffers from various effects, such as the program/erase cycle effect, data retention effect, read and program disturb, as well as cell-to-cell interference effect. In this chapter, we have discussed the characteristics of these effects. These effects cause a voltage shift, i.e. the default threshold is no longer optimal, which leads to higher error probability and asymmetric error characteristic.

A read-retry mechanism can reduce the BER. This mechanism requires a larger number of read-retry operations to adjust the threshold voltage. Based on the meta-data protected by the ECC unit, we proposed a strategy that can minimize the number of read-retries. The numerical results demonstrate that the read-retry strategy reduces the BER compared with fixed read thresholds and obtain error rates close to the results with the optimal thresholds. This strategy shows that the meta-data of one page is sufficent to compensate voltage shifts.

3 Source coding schemes for flash memories

In this chapter, different lossless data compression schemes are presented for application in non-volatile flash memories. The objective of the lossless compression scheme is to reduce the amount of redundancy of the user data without losing the original data.

Data compression is an integral part of many storage systems. It can improve the storage capacity, i.e. increase the number of data sectors that can be written per flash memory page. [28, 29, 30, 31][32, 33]. Furthermore, the flash endurance can be improved, i.e. the errors under P/E cycling stress can be reduced.

In this chapter, we propose adaptive data compression algorithms for short data blocks. The proposed compression scheme combines a modified move-to-front (MTF) algorithm [34, 35, 36] with Huffman coding [37], named the MTF-Huffman (MH) scheme. The MTF algorithm transforms the probability distribution of the input symbols to a new output distribution. A similar algorithm was proposed in [38] using adaptive Huffman coding. Omitting the adaptation of the Huffman algorithm reduces the complexity of the algorithm. We demonstrate that a Huffman code based on a discrete approximation of the log-normal distribution provides sufficient data compression for the well-known test data.

Next, we propose a modification of the MTF algorithm. The basic concept of the MTF algorithm is that each symbol in the data is replaced by its index in a stack of recently-used symbols. The stack is then re-ordered, where the current symbol is placed in the topmost position. There exists a number of modifications to this procedure, e.g. a method proposed in [39], where the current symbol is placed in the second rank. After two successive occurrences, a symbol is moved to the top position. We refer to this algorithm as MTF-1. We proposed an MTF-n algorithm, where the idea of the MTF-1 algorithm is adapted to small package sizes by placing a symbol in position n if it has a high rank. Thus, infrequent symbols do not interfere with the ranking of frequent symbols.

Furthermore, we propose an improvement for the MH scheme, which is named the BWT-MTF-Huffman (BMH) scheme. The BMH scheme is a combination of Burrows-Wheeler transform (BWT) [40], followed by MH. The objective of the BWT algorithm is to reorder the symbols to obtain a new output that is more suitable for compression. The BMH and the MH algorithms are applicable for flash memory systems where the data compression is performed at the block level with short block sizes. For a block size of 1 KB, the MH scheme provides compression performance comparable with the Lempel-Ziv-Welch (LZW) algorithm [41], as well as with the parallel dictionary LZW (PDLZW) algorithm, which is suitable for fast hardware implementations [42].

In section 3.1, we present the motivation for using data compression in flash memory. In section 3.2, we propose the encoding and decoding structure of the data compression schemes. In section 3.3, we propose a modification of the MTF algorithm. In section 3.4, we propose the MH scheme and compare the proposed scheme with the LZW algorithm [41] and the PDLZW algorithm. In section 3.5, we propose the BMH scheme, which provides a higher compression gain result. The comparison results are presented in the same section. In section 3.6, we discuss the WA results. Finally, we summarize this chapter in section 3.7. Parts of this chapter are published in [32, 33, 2, 43, 1].

3.1 Motivation

Data compression is less frequently applied for flash memories than ECC. Nevertheless, data compression can be an important ingredient in a non-volatile storage system, which improves the system reliability. For instance, data compression can reduce WA [28] due to the mismatch between erase and program operation of the flash. WA is a measure of storage efficiency. A lower WA improves the overall life expectancy of NAND flash memory. The WA can be calculated as the ratio between the amount of data written to the flash memory and the amount of data written by the host, i.e.

$$WA = \frac{\text{data written to the flash memory}}{\text{data written by the host}}. \tag{3.1}$$

The higher the WA, the more P/E cycles that are consumed writing a certain amount of user data, which leads to faster wear-out of the flash memory. Consequently, a large WA value shortens the lifetime of flash memories. WA happens due to different procedures such as garbage collection. Here, WA refers to the fact that the amount of data written to the flash memory is typically a multiple of the amount intended to be written.

Flash memory must be erased before it can be rewritten. The granularity of the erase operation is typically much larger than that of the program operation. Hence, the erase process results in rewriting user data. Data compression reduces the amount of data transferred from the host system and hence can reduce the WA and increase the logical capacity [28].

On the other hand, data compression can improve the ECC capability by introducing more redundant bits. Accordingly, a reduction of the number of information bits enables more parity bits for error correction [29] [32, 33]. As error rates continue to grow, it is necessary to increased the error correction capability of the ECC. In the next chapter, we use the data compression scheme to improve the error correction capability of the ECC unit.

Data compression schemes aim to minimize the average number of bytes per data block. Therefore, the average block length is used as a performance measure for data compression algorithms. However, if we use data compression to improve the ECC, we are interested in the minimum compression gain for data blocks, because this minimum compression gain limits the number of bits that can be used as additional parity bits. Hence, the reliability of the error correction can be improved.

3.2 Proposed data compression algorithm

A source code is universal if it can be constructed without knowledge of the statistics of the source. Frans Willems [36] proposed the universal source code based on fixed-to-variable length. The Elias-Willems source coding scheme comprises an L-block Parser, the second Elias code $B_2(n)$ and MTF for block length L messages from a discrete stationary source [36].

Figure 3.1 illustrates the essentials of the Elias-Willems source coding scheme. In the first step, the MTF algorithm is applied in a source coding scheme where

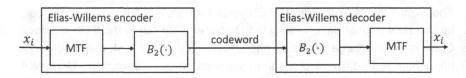

Figure 3.1: The Elias-Willems source coding scheme.

a message symbol is mapped to a codeword of variable length. The i-th codeword is selected for the current source symbol if i different symbols occurred since the last appearance of the current source symbol. The integer i is encoded to a codeword from a finite set of codewords of different lengths. In order to keep track of the recency of the source symbols, the symbols are stored in a list, ordered according to the occurrence of the symbols. Source symbols that occur frequently remain close to the first position of the list, whereas more infrequent symbols will be shifted towards the end of the list. Consequently, the probability distribution of the output of the MTF algorithm tends to be a decreasing function of the index. In order to compress the data, the indices are mapped to codewords of variable lengths, where lower values correspond with shorter codewords.

Second, Peter Elias [44, 35] introduced prefix-free binary source coding of positive integers. In this method, a positive integer n is represented by a binary sequence $B(n)$ of length $L(n) = \lfloor log_2 n \rfloor + 1$, which is not prefix-free [45]. This code can be prefix-free by adding a prefix of $L(n) - 1$ zeros to the standard codeword $B(n)$ to form the codeword $B_1(n)$. The only problem is that the codewords of $B_1(n)$ are about twice as long as those of $B(n)$, i.e. $L_1(n) = 2 \lfloor log_2 n \rfloor + 1$. In order to avoid this problem, Elias described the second Elias code $B_2(n)$, which is a combination of $B(n)$ and $B_1(n)$. The codeword of n is built by appending the prefix $B_1(L(n))$ to the codeword from $B(n)$ with the leading 1 removed. Since the code $B_1(n)$ is prefix-free and the remaining length of the codeword is uniquely determined, the second Elias code is definitely prefix-free with $L_2(n) = \lfloor log_2 n \rfloor + 2 \lfloor log_2 (\lfloor log_2 n \rfloor + 1) \rfloor + 1$. Note that the term $log_2(log_2 n)$ becomes negligible compared with $log_2 n$ for larger positive integers n. The Elias-Willems scheme is asymptotically optimal as L goes to infinity [36, 45]. However, the term $log_2(log_2 n)$ has a influence on small positive integers n, which reduces the compression performance and the coding efficiency. Therefore, this universal code may not be applicable for flash memory that works in small block size.

Data compression for flash memories has a number of constraints compared with data compression at the file level. In flash memory, the data compression has to compress small chunks of user data, because blocks might be read independently. The controller unit for a flash memory operates at a block level with typical block sizes of 512 bytes up to 4 KB. For flash memory, it is important to consider compression performance and implementation cost. Furthermore, flash memory systems typically have to provide high data rates and low latency for random access. Consequently, the data compression should not compromise the system performance. We propose an adaptive data compression scheme that considers all of the mentioned constraints.

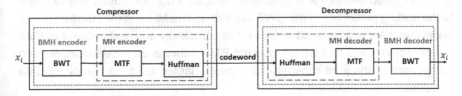

Figure 3.2: The MH and BMH schemes structure.

For small block size, we propose a lossless data compression scheme comprising two stages: an MTF algorithm followed by Huffman coding. Note that the MTF algorithm does not require a large stack to store the symbols. Figure 3.2 depicts the structure of the proposed MH and BMH encoder and decoder. For the second coding step, we use Huffman encoding, which was developed in 1952 by David Huffman [37]. Note that the Huffman code also considers as optimal prefix-free codes. The Huffman code is not an universal code, i.e. it requires knowing the probabilities of the source symbols to achieve the best possible compression ratio. Compression algorithms operating at the file level typically store the coding table with each encoded file. For the encoding of short data blocks, the overhead for such a table would be too costly. In order to avoid the table, we propose a fixed Huffman code constructed on a predetermined symbol distribution without any online adaptation. We estimate the probability distribution of the output of the MTF algorithm to avoid the memory of the source. This estimation enables fast data processing, low complexity, and is well suited for data compression of short data blocks, where the adaptation typically results in an overhead.

The input distribution of the MTF algorithm can be improved by introducing the BWT algorithm. The aim of the BWT algorithm is not to compress a message but rather to change the order of symbols into a form that is more amenable to compression. Likewise, we propose the BMH scheme comprising three stages: the BWT followed by MTF algorithm and then Huffman coding, as illustrated in figure 3.2. A similar coding scheme is used in the bzip2 data compression approach [46], for instance. However, bzip2 is intended to compress complete files. In order to adapt the compression algorithm to small block sizes, we estimate the output distribution of the combined BWT and MTF algorithm and use a fixed Huffman code. The BMH scheme provides a higher compression gain compared with MH and other schemes, although it requires memory due to the sorting procedure of the BWT algorithm. The MH and the BMH schemes show that there is a trade-off between the complexity and performance of data compression. Note that we will use the BMH scheme as a data compression component of the joint scheme in the next chapter 4, i.e. a combined data compassion with ECC unit.

3.3 Move-to-Front (MTF) and MTF modification

In this section, we present an overview of the MTF and MTF-1 algorithm. The modified MTF-n and MTF-n/1 achieve the best compression gain and the same complexity compared with the ordinary MTF and MTF-1.

3.3.1 MTF

MTF is a data transformation algorithm that assigns smaller codes to symbols that have appeared in the recent past. The MTF algorithm also called the recency rank calculator was introduced by Elias [35] and Willems [36], and is an efficient method to adapt the statistics of the user data. The main idea of the algorithm is to update the symbols in the list where the frequently-occurring symbols are moved to first position in the list. Each symbol is encoded by the position number where it appears in the alphabet. The operation of the MTF is demonstrated in the following example and illustrated in figure 3.3.

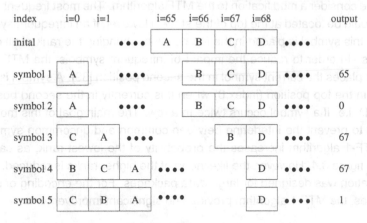

Figure 3.3: MTF for an alphabet of 256 binary symbols.

Example 1. *We consider an MTF that reads eight bits per iteration. Hence, we require a list with 256 entries. We assume that 8 bits correspond with ASCII characters and the list is initialized with the ASCII character set. Consider the input sequence X = (A,A,C,B,C,...), where the characters have the ASCII values A=65, B=66, C=67 and so on. In the first iteration, the MTF locates the symbol A in the position 65. Hence, the output of the MTF is 65. Subsequently, the value 65 is moved to the first position (index 0) of the list. In the second iteration, the symbol A is located in the first position of the list and the MTF outputs index is 0. The following iterations are illustrated in Figure 3.3.*

Source symbols that occur frequently remain close to the first position of the list, whereas more infrequent symbols will be shifted towards the end of the list. Furthermore, a symbol is encoded as the current index of that symbol in the list. Thus, the indices are mapped to codewords with variable lengths. In order to enable a fast encoding, e.g. encoding of one symbol per clock cycle, the search for the index is implemented in parallel.

3.3.2 MTF-1

Next, we consider a modification to the MTF algorithm. The most frequent symbols should be located at the top of the stack. However, if an infrequent symbol occurs, this symbol is placed in the top position, changing the rank of all other symbols. In order to reduce the impact of infrequent symbols, the MTF-1 algorithm places the current symbol in the second position [39]. A symbol is only placed in the top position (index 0) when it is currently in the second position (index 1), i.e. if a symbol occurs twice in a row. The main goal of this modification is to prevent the interfering between common and uncommon symbols. The MTF-1 algorithm increases the probability of the lowest rank, as can be seen in figure 3.4. However, the likelihood of the higher ranks is reduced. This modification was designed for large data packages. For the encoding of short packages, the MTF-1 algorithm provides no significant improvements.

3.3.3 MTF-n

In order to optimize the MTF algorithm, we propose a simple modification, namely the MTF-n algorithm. After a certain number of iterations, the symbols in the stack are ordered according to their likelihood. Symbols with a low rank have a high probability, whereas higher ranks indicate lower probabilities.

Similar to the MTF-1 algorithm, we prevent a symbol with low rank and presumably low probability from being placed in the top position of the stack. We use two parameters n and $m \geq n$ to control the MTF operation. If the current rank of a symbol is greater than or equal to m, this symbol is placed in position n of the stack, otherwise it is placed in the top position. The processing results in a hysteresis ensuring that a symbol occurring in close succession is moved to the front of the stack, whereas infrequent symbols will not obtain a rank lower than n. The best values of n and m are 9 and 13, respectively.

Figure 3.4 depicts the output distribution of different MTF algorithms. By considering MTF-n with $n = 9$ and $m = 13$, the probability for ranks 2 to 7 are increased compared with MTF and MTF-1. The likelihoods for larger ranks are reduced, but on average the compression gain is improved. Furthermore, it can be observed from figure 3.4 that MTF-n reduces the likelihood for the first

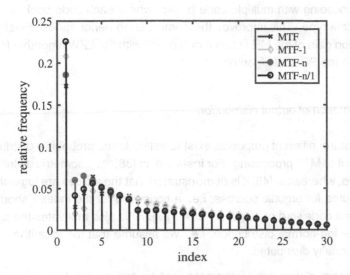

Figure 3.4: Distributions of index values for different versions of the MTF algorithm.

rank compared with the MTF-1 algorithm. This can be omitted by a combination of both algorithms, which we denote by MTF-n/1. If the current rank of a symbol is greater than or equal to m, this symbol is placed in position n of the stack, otherwise it is placed in the second position. A symbol is only placed in the top position when it is currently in the second position, i.e. if a symbol with a rank smaller then m occurs twice in a row. We will see in section 3.4 that this algorithm achieves the highest probability for rank one and the best compression gain.

3.4 MTF-Huffman (MH) scheme

In this section, we investigate different probability distributions to estimate the probability distribution at the output of the MTF encoding. In particular, we consider the Calgary and Canterbury corpora [47, 48]. Both corpora include real-world test files to evaluate lossless compression methods. We observe that both corpora data content different types of files. Therefore, we propose

Huffman encoding with multiple code books, where each code book contents a similar data type. This improves the compression performance. Finally, the compression gain of the MH scheme compares with the LZW algorithm [41] as well as with the PDLZW algorithm.

3.4.1 Estimation of output distribution

In the literature, different proposals exist to estimate the probability distribution of the output of MTF processing. For instance, in [38] the geometric distribution is proposed, whereas in [49] it is demonstrated that the indices are logarithmically distributed for ergodic sources, i.e. a codeword for the index i should be mapped to a codeword of length $L_i \approx \log_2(i)$. This also motivates the application of the log-normal distribution, i.e. we assume that the logarithm of the index is normally distributed.

The log-normal distribution is characterized by two parameters, the mean μ and the standard deviation σ. The probability density function for the log-normally distributed positive random variable x is

$$p(x) = \frac{1}{\sqrt{2\pi}\sigma x}\exp\left(-\frac{(\ln(x) - \mu)^2}{2\sigma^2}\right). \qquad (3.2)$$

For the integers $i \in \{1, \ldots, N\}$, we propose a discrete approximation of the log-normal distribution. We obtain the discrete probability distribution

$$P(i) = \frac{p(\beta i)}{\sum_{i=1}^{N} p(\beta i)}, \qquad (3.3)$$

where β denotes a scaling factor. The μ, σ, and β can be adjusted to approximate the probability distribution at the output of the MTF for a real-world data model. Hence, we calculate the Huffman code once based on the predetermined probability distribution.

In order to fit the parameters of the distribution, we use the Kullback-Leibler (KL) divergence as a performance measure. The KL divergence is a non-symmetric measure of the difference between two probability distributions. Let

$Q(i)$ and $P(i)$ be two probability distributions. The KL divergence is defined as [50]

$$D(Q\|P) = \sum_i Q(i) \log_2 \frac{Q(i)}{P(i)}. \tag{3.4}$$

The KL divergence is always non-negative and a smaller value corresponds to a better approximation. The KL divergence represents the expected number of extra bits per input symbol that must be transmitted if a code is used corresponding to the probability distribution $P(i)$ instead of the actual distribution $Q(i)$.

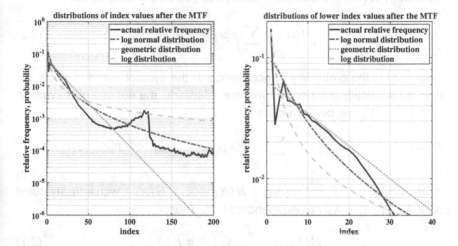

Figure 3.5: Distributions of index values after the MTF algorithm for MH.

Figure 3.5 depicts the different probability distributions as well as the relative frequencies for the Calgary corpus. The blue curve corresponds to the relative frequencies of index values averaged over all files from the Calgary corpus. The other curves are the best matches of the considered distributions obtained by minimizing the KL divergence. The figure on the left side shows the complete output distribution after MTF. As we can see, the peak of the relative frequencies occurs due to different data types. The indecies with high probability are more important to increase the compression gain. Therefore, we present the right figure, which illustrates the distribution for lower indecies. This depicts that the relative frequencies are well approximated by the log-normal distribution.

Additionally, we present results for all presented approximations in Table 3.1. The first two rows in the table present values for the KL divergence. All distributions are compared with the actual output distribution of the MTF processing. Table 3.1 presents the average block length in bytes per 1 KB block, which is denoted by $mean$. In the second column, it is an upper page (UP) on the entropy. Shannon's entropy was introduced in [51]. It is used to calculate the minimum number of bits on average that is necessary per message symbol. The entropy $H(X)$ of a discrete memoryless source x is defined as

$$H(X) = \sum_{x \in X} p(x) \, log_2 \frac{1}{p(x)}, \tag{3.5}$$

where $p(x)$ is the probability of occurrence of the symbol x and symbols occur in a statistically independent manner. Similarly, the entropy of n symbols $X_1, X_2, \cdots, X_n$ can be defined as

$$H(X_1, X_2, \ldots, X_n) = - \sum_{x_1, x_2, \ldots, x_n} p(x_1, x_2, \ldots, x_n) \, log_2 \, p(x_1, x_2, \ldots, x_n). \tag{3.6}$$

For $H(X_1, X_2, \ldots, X_n) \leq H(X_1) + H(X_2) +, \ldots, H(X_n)$, with equality if and only if $X_1, X_2, \cdots, X_n$ are independent [52]. Thus,

$$H(X_1, X_2, \ldots, X_n) \leq n \, H(X). \tag{3.7}$$

Based on equation 3.7, we calculate the upper bound (UB) in the second column.

For different probability distributions, all parameters were optimized for the corpus using data blocks of 1 KB and a list of length $N = 256$. The list is initialized after each data block. In the third column in Table 3.1, the log-normal distribution is optimized with $\beta = 1$ and the best parameters are $\mu = 1$, and $\sigma = 4.5$, which minimizes the average block length, the maximum block length and KL divergence. By introducing β to the log-normal distribution, the best parameters are ($\mu = 1.2$, $\sigma = 1.7$, and $\beta = 0.4$), which optimizes the KL divergence for both corpora. Note that these parameters are optimal among all files but not for the obj files.

Table 3.1: Results for different probability distributions. The average block length in bytes per 1KB block is denoted by $mean$, whereas $\overline{max}$ is the maximum block length per file in bytes, averaged over all files in the corpus.

	(UB)	log-normal (1,4.5)	log-normal (1.2,1.7,0.4)	geometric $p = 0.06$	log. dist.
$D(Q\|P)$ Calgary	–	0.210	0.115	0.225	0.260
$D(Q\|P)$ Canterb.	–	0.25	0.218	0.462	0.321
$mean$ Calgary	583.7	713.9	689.1	714.3	722.8
$mean$ Canterb.	463.1	626.1	623.7	666.3	644.5
$\overline{max}$ Calgary	732.6	822.7	803.1	850	824.1
$\overline{max}$ Canterb.	679.8	770.7	746.6	769	774.9

The log-normal distribution leads to the highest compression gain regarding to $mean$ and $\overline{max}$ for both corpora. Note that all results for the Canterbury corpus are obtained using the same parameters optimized for the Calgary corpus. The geometric distribution minimizes the average block length, the maximum block length and KL divergence with $p = 0.06$.

Typically, the average block length $mean$ is considered as a performance measure for the WA in flash memories. However, if we use the compression scheme to improve the error correction capability in the ECC unit, then we consider the maximum block length max to determine the extra additional parity bits for the ECC unit. The two bottom rows in Table 3.1 present the maximum length $\overline{max}$ for all data blocks within a single file of the corresponding corpus, where the values are averaged over all files in the corpus. The log-normal distribution also results in the smallest maximum length for both corpora.

3.4.2 Huffman encoding with multiple code books

Based on the investigation of Calgary and Canterbury corpora, the corpora data have different types of files; for example, text files, object files, and picture files. We demonstrate that the encoding can be adapted to different data types by using several code books for the Huffman encoding. Each code book can be approximated based on the log-normal distribution, but with different values for the μ, σ, and β.

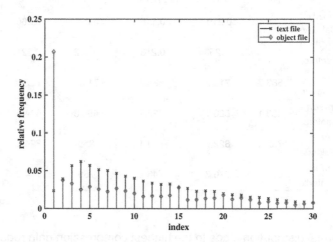

Figure 3.6: Distributions of index values after the MTF algorithm.

Figure 3.6 presents two examples of distributions at the output of the MTF. The distributions correspond to a text file and an object file from the Calgary corpus. For the object file, the first index indicates a higher relative frequency compared with the text file. These distributions can be considered as extreme cases because μ value and the σ show the largest differences among the different file types.

In [32], a distribution with $\mu = 1$ and $\sigma = 4.5$ was proposed as a compromise for different file types. In Table 3.2, we use a new set of parameters, which is applicable for text files but not for the other files. As can be observed in Table 3.2, there is no optimal log-normal distribution for all files. For example, the *paper1* has better compression gain by using the first set of parameters, although using the second set is more suitable for the *obj1*. Therefore, we use one set of parameters, which is a compromise for all files and achieves the best gain among different types. A simple adaptation to different data types can be achieved by using several code books for the Huffman encoding for different μ,

σ, and β values, where two code books already achieve a compression gain. For the MTF with two code books, we add one bit of media data to select which Huffman code is used.

Table 3.2: MTF compression gains for different log-normal distributions.

[Files names]	Entropy (UB)		($\mu = 1.2$, $\sigma = 1.7$, $\alpha = 0.4$)		($\mu = 2$,$\sigma = 3.7$,$\alpha = 0.1$)	
	mean	max	mean	max	mean	max
paper1	669.4	719.7	729.9	759.2	778.6	806.7
'obj1'	688.3	899.9	790.6	1022.5	784.6	949.5

Considering *obj1* and *paper1* from Table 3.2, the best $mean$ and max using one code book of the first parameter are 760.2 and 890.3 bytes on average, respectively, whereas using two code books improves the gain performance for the $mean$ and max by 757.2 and 854.3 bytes on average, respectively. This illustrates that the two code books can achieve a good compression perform- ance, although it requires two encoders and has higher latency.

3.4.3 Results for MH scheme

In this sub-section, we compare the compression gain of the MH scheme with the LZW algorithm [41], which is widely used for lossless data compression. This scheme is universal code developed by Ziv and Lempel where the com- pression method is based on a dictionary method. The dictionary method chooses strings of symbols and encode each string as token using a dic- tionary, i.e. each string is mapped to their position in the dictionary. Ziv and Lempel [53] introduced LZ77 and LZ78 algorithms, where the main dif- ference between these algorithms is how the dictionary is built. Subsequently, Welch [41] developed an LZW algorithm based on LZ78. The LZW algorithm constructs a code book of sequences encountered in the data as it is encoded, where the code book contains sequences of different length that are stored recursively in a single dictionary. Several concepts have been proposed to improve the throughput of a hardware implementation of the LZW algorithm. For instance, hashing and tree-based search techniques lead to a fast LZW algorithm [54]. Some other algorithms are the DLZW (dynamic LZW), WDLZW (word-based DLZW) [55], and PDLZW (parallel dictionary LZW). The PDLZW algorithm uses several code books, each of which contains only sequences of the same length [56, 42], i.e. a hierarchical set of parallel code books is used with increasing word widths. This architecture is well suited for fast VLSI implementations due to its high regularity. It provides faster compression and decompression compared with the standard LZW algorithm because it does not need to search the dictionary recursively as conventional implementations. The proposed data compression algorithm is compared with the original LZW algorithm as well as the PDLZW algorithm, where we use code books with 1024 entries for both versions.

Table 3.3: Results for different MTF compression algorithms. The average block length in bytes per 1KB block is denoted by $mean$, whereas $\overline{max}$ is the maximum block length per file in bytes, averaged over all files in the corpus.

Corpus names	Entropy	MH scheme (log-normal distribution)				LZW		
	UB	MTF	MTF-1	MTF-n	MTF-n/1	LZW	PDLZW 4 dic.	PDLZW 7 dic.
mean Calgary	583.7	689.1	685.0	672.5	667.2	649.3	691.3	677.4
mean Canterbury	447.5	623.1	613.7	608.7	599.4	470.3	561.8	507.8
$\overline{max}$ Calgary	732.6	803.1	799.7	787.0	787.3	816.6	853.6	860.1
$\overline{max}$ Canterbury	668.9	746.6	742.2	733.1	727.0	730.2	759.2	734

The compression results with respect to the compression gains for the proposed algorithms as well as for both versions of the LZW algorithm are presented in Table 3.3. All MTF algorithms use the log-normal distribution with optimized parameter sets. For the Calgary and Canterbury corpus, the MH algorithms outperform the LZW algorithms with respect to the maximum block length, as indicated by the $\overline{max}$ values.

Regarding the average compression gains of the Calgary corpora, the LZW algorithm provides better values, whereas the performance of the MH algorithm is better than the PDLZW. The performance and complexity of the PDLZW depend on the number of parallel dictionaries, where Table 3.3 considers two implementations with four or seven dictionaries, respectively. Note that all PDLZW algorithms are unable to compress all data blocks of size 1 KB contained in the Calgary corpus, where they fail in five cases, whereas all MTF algorithms are able to compress all data blocks with the same block size.

For the Canterbury corpora, the average compression gains of the LZW outperform the MH algorithms. The PDLZW algorithms using four or seven dictionaries have a better performance compared with the MH algorithms. Note that all considered algorithms are able to compress all data blocks of size 1KB that are contained in the Canterbury corpus without any failure.

3.5 BWT-MTF-Huffman (BMH) scheme

In this section, we propose the BMH scheme, which a BWT as well as a combination of MTF and Huffman coding. This scheme improves the compression performance of the MH scheme by considering successive processing starting with the BWT and then with the MH. This designed improves the coherence in the data [40]. Furthermore, it arranges the data input in such a way that the transformed message is more compressible. Hence, a better compression rate is achievable by combining the BWT and the MTF. The Huffman code is constructed from an estimated output distribution of the combined BWT and MTF algorithms. In order to fit the estimated distribution to the relative frequencies, we propose a parametric logarithmic distribution. Finally, the proposed BMH algorithm is compared with the original LZW algorithm and the PDLZW algorithm.

3.5.1 Burrows-Wheeler Transform (BWT)

The BWT is a block sorting algorithm that is applied to sequences with fixed length to produce a permutation for the input data [40]. Let S be an input sequence that includes N symbols. Subsequently, the output of the BWT(S) permutes the symbols of the same length. However, the data in the output sequence are sorted lexicographically. A major drawback of the BWT algorithm is its high complexity, i.e. the algorithm has time and space complexity of order $O(N^2)$. In order to reduce the complexity of the block sorting, we used modified sorting using a limited context order k of sorting [57, 39]. We consider $k = 4$ for the proposed algorithm. The limited context reduces the complexity to $O(kN)$, where $k < N$.

Let the string $S = S[0, \cdots, N-1]$ be the input string of length N. The BWT comprises four basic steps:

- Append to the end of S a special symbol "$"that is used to uniquely indicate the last symbol of the sequence.
- The cyclic shift of order i of the string S is $S^{[i]} = [S[i, \cdots, N], S[0, \cdots, i-1]]$ for each given $0 \le i \le N$. For each cycle shift $S^{[i]}$ is stored in an ascending order in a matrix M.
- The conceptual $N \times N$ matrix M whose rows are sorted lexicographically.
- Construct the transformed string $\hat{S} = \text{BWT}(S)$ by taking the last column l of the matrix M, includes an index j which indicates the position of the first symbol of the original sequence.

The inverse of $\hat{S}$ can be obtained from the last column l and first column f of the matrix M, including the index j (see [58] for details).

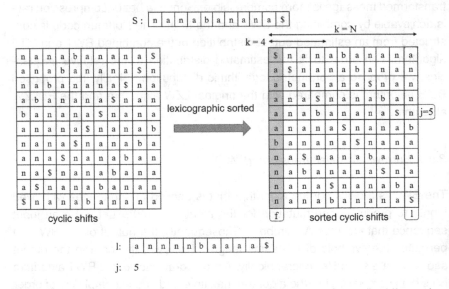

Figure 3.7: BWT encoding of matrix M, left matrix with successive cyclic shifts, and right matrix with a full context depth sorting order

Example 2. *Figure 3.7 illustrates the BWT encoding process for full context sorting $k = N$. Let the symbol sequence S ="nanabnana\$" is stored in the matrix M. The following rows of M are obtained by successive cyclic shifts of the first row. Afterward, the rows of the matrix M are sorted in lexicographic order. The output of the BWT transform is BWT(S)= "annnnbaaaa\$" which is the last column l of M includes the index j of the symbol n which is equal to 5.*

Example 2 shows that the probability of finding the same character close to another is substantially increased. Clearly sorting the whole matrix achieves better compression rate, but we may sort this matrix up to the order $k < N$ to reduce the complexity. Furthermore, the encoding has a small compression loss and a slower inverse transformation [58]. However, this limited context order BWT is suitable for hardware implementations. For instance, BWT encoder architectures were proposed in [59, 60, 61, 62, 63]. The BWT encoders achieve high data throughput due to parallelization of the sorting. In [58], a pipelined decoder architecture for the limited context BWT is proposed, which comprises $O(k)$ pipeline stages. This results in a fast decoding operation and a high data throughput. The complexity of the pipelined decoder is relatively low, because the size of the logic is only of order $O(N)$ and the memory requirements of order $O(kN)$. However, the pipeline requires a latency of order $O(kN)$ clock cycles.

3.5.2 Estimation of output distribution for BMH

In order to adapt the estimation of the output distribution to the two-stage processing of BWT and MTF, we introduce a modification of the logarithmic distribution as proposed in [49]. For any integer $i \in \{1, \ldots, N\}$, the logarithmic probability distribution $P(i)$ is defined as

$$P(i) = \frac{1}{i \sum_{j=1}^{N} \frac{1}{j}}. \tag{3.8}$$

The logarithmic distribution depends only on the number of symbols N.

Now consider the BMH scheme. With the BWT each symbol keeps its value but the order of symbols is changed. If the original string at the input of the BWT contains substrings that often occur, then the transformed string will have

several places where a single character is repeated multiple times in a row.
For the MTF algorithm, these repeated occurrences result in sequences of
output integers all equal to 1. Consequently, applying the BWT before the MTF
algorithm changes the probability of rank 1. In order to take the BWT into
account, we propose a parametric logarithmic probability distribution

$$P(1) \;=\; P_1$$
$$P(i) \;=\; \frac{1}{i(P_1 + \sum_{j=2}^{N} \frac{1}{j})} \text{ for } i \in \{2, \ldots, N\}. \tag{3.9}$$

Note that with the ordinary logarithmic distribution we have $P_1 \approx 0.1633$ for
$N = 256$. With the parametric logarithmic distribution, the parameter P_1 is the
probability of rank 1 at the output of the cascade of BWT and MTF. Therefore,
we estimate P_1 according to the relative frequencies at the output of the MTF
for the Calgary and Canterbury corpora. We use only the Calgary corpus to
determine the value of P_1, which results in $P_1 = 0.4$. Note that the Huffman
code is not very sensitive to the actual value of P_1, i.e. for $N = 256$ values in
the range $0.37 \le P_1 \le 0.5$ result in the same code.

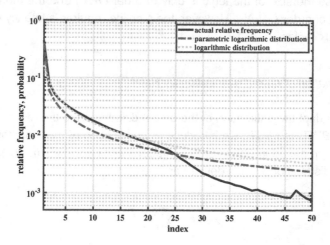

Figure 3.8: Distributions of index values after BWT-MTF algorithm for the actual relative frequency, parametric log distribution, and the log distribution.

Table 3.4: KL divergence for the actual output distribution of the BMH processing and the approximations for all files of the Calgary corpus.

file	log-dist.	parametric log. dist.
trans	0.539	0.195
progp	0.700	0.276
progl	0.713	0.314
progc	0.486	0.207
pic	1.773	0.827
paper6	0.455	0.264
paper5	0.436	0.266
paper4	0.467	0.346
paper3	0.454	0.367
paper2	0.477	0.363
paper1	0.427	0.273
obj2	0.559	0.125
obj1	0.375	0.045
news	0.321	0.239
geo	0.160	0.046
book2	0.456	0.320
book1	0.454	0.447
bib	0.377	0.200

Figure 3.8 depicts the index values after BWT followed by MTF algorithm for the parametric logarithmic, the logarithmic distribution, and the relative frequencies for the Calgary corpus. Note that the compression gain is mainly determined by the probabilities of the low index values. Table 3.4 presents values for the KL divergence for the log distribution and the proposed parametric log distribution with $P_1 = 0.4$. Both distributions are compared with the relative frequencies of the BWT followed by MTF processing, where both transformations are initialized after each data block. Note that the proposed parametric distribution results in smaller values of the KL divergence for all files in the corpus. These values can be interpreted as the expected extra number of bits per information byte that must be stored if a Huffman code is used.

3.5.3 BMH results

In this sub-section, we compare the compression gain of the BMH algorithm with other algorithm. Table 3.5 presents results for the average block length for different probability distributions and compression algorithm. All results present the average block length in bytes and were obtained by encoding data blocks of 1 KB, where we used all files from the Calgary corpus.

Table 3.5: Detailed results for the Calgary corpus for the compression of 1 KB data blocks. The mean values are the average block length in bytes whereas the maximum values are the worst-case compression result for each file.

	BMH		MH		LZW	
	parametric log. dist.		$\mu = 1.2$ & $\sigma = 1.7$ & $\alpha = 0.4$			
file	mean	maximum	mean	maximum	mean	maximum
trans	508.0	660.9	746.0	820.2	701.7	818.8
progp	442.3	607.5	710.8	760.3	634.2	755.0
progl	447.3	565.6	681.7	753.2	632.3	726.25
progc	530.9	624.6	749.9	790.8	714.0	800.0
pic	218.4	584.5	441.9	674.6	201.4	687.5
paper6	557.3	623.0	715.4	774.2	719.4	790
paper5	569.5	606.1	723.1	754.3	737.2	787.5
paper4	580.4	644.1	710.4	769.7	726.0	775
paper3	598.5	651.1	714.1	743.5	734.4	778.8
paper2	583.1	652.3	711.9	758.3	720.6	792.5
paper1	577.5	658.1	729.4	759.1	734.2	795
obj2	495.3	908.3	832.8	979.2	684.7	1001.3
obj1	580.7	930.5	790.6	1022.5	716.4	1010.0
news	634.4	738.0	746.6	812.3	790.7	883.8
geo	747.6	799.3	857.7	904.5	856.3	907.5
book2	575.9	656.0	712.4	806.1	725.7	795.0
book1	626.6	677.1	707.2	737.3	739.0	788.8
bib	583.9	635.0	786.5	809.5	771.3	797.5

The results of the BMH algorithm are compared with the LZW algorithm [41] and MH scheme. Moreover, the maximum values indicate the worst-case compression result for each file, i.e. these maximum values indicate how much redundancy can be added for ECC. Note that the BMH algorithm outperforms the LZW as well as the MH approach for almost all input files. Only for the image file named *pic* does the LZW algorithm achieve a better mean value, but not in the maximum value.

Table 3.6: Results for three stages compression algorithms. The average block length in bytes per 1KB block is denoted by $mean$, whereas $\overline{max}$ is the maximum block length per file in bytes, averaged over all files in the corpus.

Corpus names	Entropy UB	BMH scheme	
		parametric log. dist.	log. dist
mean Calgary	466.6	529.7	590.0
mean Canterbury	318.9	396.2	522.7
$\overline{max}$ Calgary	613.2	679	680.8
$\overline{max}$ Canterbury	533.6	582.9	621.2

Table 3.6 summarizes results for the complete corpus, where the values are averaged over all files. The maximum values are also averaged over all files. The results of the third and the fourth columns correspond to the BMH compression scheme using two different estimations for the output probability distribution. The third column corresponds to the results with the proposed parametric distribution, where the parameter was obtained using data from the Calgary corpus. The parametric distribution leads to a better $mean$ and $\overline{max}$ value compared with the log distribution. Note that the parametric distribution achieves better compression gain compared with the log-normal distribution. Furthermore, the BMH is compared with the LZW algorithm as well as to the PDLZW algorithm in Table 3.3. Note that the BMH algorithm achieves significant gains compared with these approaches regarding $mean$ and $\overline{max}$ value.

3.6 Write amplification (WA) results

In this section, we consider MH and BMH schemes to reduce the WA in flash memory. Nowadays, the controller unit for flash memory operates on block size 1 KB, 2 KB or 4 KB. From the previous discussion, we consider compression schemes with a block size of 1 KB. On average, the BMH reduces WA by around 50% and even more. While the MH algorithm has almost the same performances compared with LZW and PDLZW.

Table 3.7: Results for BMH and MH schemes. The average block length in bytes per 4 KB block is denoted by $mean$, and compression gain of 1,2,3 KB.

	BMH (used 4 KB)				
	1 KB	2 KB	3 KB	4 KB	**Mean (797 blocks)**
percentage %	11	45	43	1	**55**
	BMH (used 1 KB for BWT and 4 KB for MH)				
percentage %	11	18	71	0	**49**
	MH (used 4 KB)				
percentage %	0	14	76	10	**37**
	LZW (used 4 KB)				
percentage %	12	9	78	1	**45**
	PDLZW (used 4 KB, 4 dict.)				
percentage %	0	16	71	13	**38**
	PDLZW (used 4 KB, 7 dict.)				
percentage %	9	9	71	11	**41**

In order to improve the efficiency of the flash memory, we consider the block size with 4 KB. The objective of the data compression algorithm is minimizing the number of blocks that has to be operated by the flash controller or the ECC unit. Note that the compression gain is based on block units. Table 3.7 demonstrates different compression schemes, where 797 blocks are obtained from Calgary corpus. This table presents the percentage of the number of blocks which are compressed to 3 KB or less. In other words, the compression gain of block size with 4 KB are 1, 2, 3 KB, or no gain, which are $25\%, 50\%, 75\%$ or no gain, respectively. The BMH scheme achieves the highest compression gain where 56% of the blocks are compressed to the half block size or less.

Table 3.7 depicts that the BMH scheme can compress 43% of the blocks with size 4 KB to 3 KB, but 1% of the blocks are failed.

In BMH scheme, BWT has high latency and space complexity by processing such large blocks i.e. 4 KB. This complexity has an impact on the hardware implementation [58]. In order to reduce the complexity, we reduce the block sorting size of the BWT to 1 KB. Hence, we have at the output of the BWT a 4 KB block size that is sorted only 1 KB separately. Thus, we process the output of 4 KB to the MTF followed by Huffman code. However, the complexity/performance ratio becomes better, because the hardware implementation requires only approximately 25% of the hardware resources, and the compression performance slightly reduce to 49% on average as illustrated in Table 3.7. Using this method, 71% of the blocks are compressed to 3 KB, and the rest to the half block size or less. We note that the MH scheme have approximately 76% and 14% of the blocks which are compressed to 3 KB and 2 KB, respectively, but 10% of the blocks have no compression gain.

On the other hand, the LZW achieves compression gain about 45% on average. This algorithm has a high complexity with respect to hardware implementation. The PDLZW algorithms using 4 or 7 dictionaries have low compression performance compared with LZW, but the complexity is lower. Finally, both versions of the proposed BMH algorithm achieve the best compression performance compared with LZW and PDLZW. The MH scheme has a very low complexity, low latency, high data rate and a comparable performance compared with PDLZW.

3.7 Summary

In this chapter, we have presented a data compression algorithm for short data blocks. The compression is based on the MTF algorithm followed by Huffman coding, i.e. MH scheme. The proposed MTF-n and MTF-n/1 versions provide significant compression gains compared with the standard MTF algorithm. In the MH scheme, the Huffman code is constructed from a discrete approximation of the log-normal distribution. However, there is no optimal log-normal distribution for all file types. A simple adaptation to different data types can be achieved by using several code books for the Huffman encoding for different mean values and standard deviations, where two code books already achieve a significant compression gain and require only one additional bit of information per data block. However, it has high latency and requires two encoders.

Furthermore, we have presented an improvement of MH scheme, where we proposed BWT algorithm followed by MTF algorithm, and then Huffman code, i.e. BMH scheme. The Huffman code is constructed from an estimated output distribution of the combined BWT and MTF processing. For this estimate, we proposed the parametric logarithmic distribution which provides the highest compression gain.

All proposed compression algorithms (i.e. MH and BMH) achieve good compression results and lower complexity compared with other algorithms. To measure the performance for these schemes, we used the average block length in bytes per 1 KB block. The BMH outperforms LZW and PDLZW algorithms regarding average block length in all corpora. For the Calgary corpus, the LZW algorithm has a better average compression ratio than the proposed MH scheme. The PDLZW algorithm is even more critical with respect to the average length compared with MH scheme.

The main goal of the proposed data compression scheme is to reduce the WA which WA adversely affects the lifetime of flash memories. Using 4 KB block size, the BMH and MH schemes reduce the WA by about 55% and 37% on average, respectively. In order to reduce the complexity of BMH scheme, we reduce the block sorting size of BWT to 1 KB and obtain the compression scheme with 4 KB. This strategy reduces the WA by around 50%.

The MH has a very low implementation complexity and memory requirements compared with BMH and PDLZW, where the hardware complexity is dominated by the MTF algorithm. Furthermore, it provides fast data encoding and decoding. The encoder processes user data in blocks of 1 KB reading one byte per cycle. Both encoding and decoding processes have very low latency and require each only 8×256 bits of memory to adapt to the statistics of the user data. In [43], the encoder and decoder architectures of MH are proposed, where the encoder and decoder structures were implemented in Verilog and verified on an field programmable gate arrays (FPGA) board using Synopsis HAPS DX7.

The encoder and decoder have similar size with the MTF as well as for the MTF-n algorithm. The size of encoder and decoder of MH scheme is compered to PDLZW in [64]. This comparison is based on the register based implementation. The MH encoder and decoder require about 21% of the number of flip-flops and only 28% of the look-up tables (LUT) compared with the PDLZW. Note the PDLZW algorithm has relative high memory requirements which dominate the costs of a hardware implementation [64, 56].

4 A Source and channel coding approach for flash memories

The introduction of MLC and TLC technologies reduced the reliability of flash memories significantly compared with SLC flash [65, 5]. With MLC and TLC flash cells the error probability varies for the different states. Hence asymmetric models are required to characterize the flash channel, e.g. the binary asymmetric channel (BAC) [66, 20, 21, 8].

In this chapter, we propose a combined channel and source coding approach that improves the reliability of MLC and TLC flash memories. The objective of the data compression algorithm in this chapter is to reduce the amount of user data such that the redundancy of the ECC can be increased. The additional redundancy improves the reliability of the data storage system. This can be achieved, for instance, with a joint source/channel approach as proposed in [30, 31], whereas we propose a disjoint approach, where the data compression scheme operates independently of the ECC.

Moreover, data compression can be utilized to exploit the asymmetry of the channel to reduce the error probability. With redundant data, the proposed combined coding scheme results in a significant improvement of the P/E cycling endurance and the data retention time of flash memories. The combined coding scheme is based on BCH codes (cf. [67, 68]) as the ECC component and BMH scheme is the data compression component.

The remainder of this chapter is structured as follows. In the next section, we outline the motivation for using data compression to support the ECC unit. The preliminaries for ECC are briefly presented in section 4.2. The joint coding scheme is proposed in section 4.3. Next, we present an analysis of the coding scheme and some numerical results based on empirical data for errors in TLC flash in section 4.4. Finally, we summarize this chapter in section 4.5. Note that the data compression algorithms are presented in chapter 3. Parts of this chapter are published in [2, 1].

© The Editor(s) (if applicable) and The Author(s), under exclusive license to
Springer Fachmedien Wiesbaden GmbH, part of Springer Nature 2020
M. Rajab, *Channel and Source Coding for Non-Volatile Flash Memories*,
Schriftenreihe der Institute für Systemdynamik (ISD) und optische
Systeme (IOS), https://doi.org/10.1007/978-3-658-28982-9_4

4.1 Motivation

In order to ensure reliable information storage, ECC is required. For instance, BCH codes are often used for error correction that can correct multiple bit errors [69, 65, 70, 71, 72]. Moreover, concatenated coding schemes were proposed, e.g. product codes (PC) [73], concatenated coding schemes based on trellis coded modulation and outer BCH or Reed-Solomon (RS) codes [69, 74, 75], GCC [76, 77] [78] and low-density parity-check (LDPC) codes [79, 80, 81] have also been proposed for ECC in NAND flash memories.

On the other hand, data compression can also be used to improve system reliability. In the previous chapter, the data compression reduces WA and hence the logical capacity is increased. Alternatively, data compression can be used to improve the error correction capability of the ECC unit. Flash memory provides a spare area that is used to store the redundancy required for the ECC. This spare area determines the code rate and the best possible level of the error correction code. Now, this level will be determined by the compression gain of the data compression algorithm as well as by the possible code rates of the ECC unit.

For instance, for a BCH code applied on 1 KB of user data, the error correction capabilities could be multiples of 12 in the range from 12 to 96 bit errors per flash sector [72]. For this coding scheme, at least 21 bytes of additional redundancy would be required to switch from one ECC level to the next. This scenario affects the objective of the data compression algorithm. Typically, a data compression scheme would be designed to minimize the average number of bytes per data block. In order to improve the reliability of the system, the objective of the data compression algorithm should be to minimize the number of data blocks that cannot be compressed sufficiently to switch to a higher error correction level than the default ECC level for the particular flash type.

Furthermore, data compression can be utilized to exploit the asymmetry of the channel. Coding schemes were proposed that take these error characteristics into account [82, 83, 84, 85, 86]. The flash channel is typically considered as a binary channel with input symbols 0 and 1 [65]. However, for MLC and TLC the probability that an input 0 will be flipped into a 1 differs from the probability for a flip from 1 to 0 [87, 88, 8]. In this chapter, we assume that the flash can be modeled as BAC where an input 0 has a lower error probability. For the

BAC, the compression gain of the data compression can be exploited by using systematic ECC encoding and zero-padding for the compressed information bits, where the zero-padding reduces the error probability.

4.2 Preliminaries for error correction coding

In this section, we provide an overview of the basic concepts and definitions of algebraic codes. We discuss the basics of the Galois fields (GF) that enable basic operations on a finite number of symbols. Subsequently, we provide an introduction o linear block codes (e.g. BCH code).

4.2.1 Finite fields

Finite fields are also called GFs. A GF is a field with a finite number of elements which provides the mathematical operations, i.e. the operations addition and multiplication are defined, while satisfying the axioms of a field [67]. The mathematical operations are necessary for algebraic codes.

Definition 1. *Prime field $GF(p)$ [67, 89]*
Let p be a prime, the set of elements $\{0, 1, 2, \ldots, p-1\}$ with the operations addition $(+)$, multiplication $(\cdot)$ mod p, satisfies the axioms of a field and is called a prime field $GF(p)$. Let α be the primitive element of $GF(p)$, such that all of the non-zero elements of the $GF(p)$ can be obtained from powers of α.

A polynomial can be expressed over GF(p) in the form $a(x) = a_0 + a_1 x + a_2 x^2 + \cdots + a_{n-1} x^{n-1}$, where the degree $n-1$ is a non-negative integer and the coefficients $a_i \in GF(p)$. The set of polynomials over $GF(p)$ is denoted as $GF(p)[x]$. In coding theory, sometimes the polynomials are expressed as vectors and vice versa, therefore the bijective relation is used as follows:

$$\mathbf{a} = (a_0, a_1, \ldots, a_{n-1}) \longleftrightarrow a(x) = a_0 + a_1 x + \ldots + a_{n-1} x^{n-1}. \quad (4.1)$$

Definition 2. *Irreducible polynomial [67]*
A polynomial $p(x)$ over $GF(p)$ is irreducible over $GF(p)$ if $p(x)$ is not divisible by any other polynomial over $GF(p)$ of degree less than the degree of $p(x)$ but greater than zero.

Definition 3. *Extension field $GF(p^m)$ [67]*
Let $p(x)$ be irreducible over $GF(p)$ and $\alpha \notin GF(p)$ be root of $p(x)$ with the degree $p(x) = m$, and with operations $(+,\cdot) \bmod p(x)$. Then, $GF(p^m)$ is the smallest field that contains $GF(p)$ and α.

Definition 4. *Primitive polynomial*
A polynomial p(x) is primitive if all powers of α generate all elements of an extension field. Primitive polynomials are also irreducible polynomials.

4.2.2 Linear block code

For linear block codes, a block of k information symbols is encoded into a block codeword of n symbols $n > k$. The code length is n and the number of redundancy symbols is $r = n - k$. The code rate is $R = \frac{k}{n}$.

Definition 5. *Linear block code [90]*
A code $\mathcal{C} \subset GF(p^m)^n$ is a linear code if for any two codewords $\mathbf{c_1}, \mathbf{c_2} \in \mathcal{C}$ the following is satisfied:

$$a\mathbf{c_1} + b\mathbf{c_2} \in \mathcal{C}, \quad a, b \in GF(p^m) \tag{4.2}$$

Definition 6. *Hamming weight [67]*
For a given $c \in GF(p^m)^n$, the Hamming weight is the number of non-zero elements in c. For a symbol c_j, the Hamming weight is:

$$wt_H(c_j) = \begin{cases} 0, c_j = 0 \\ 1, c_j \neq 0 \end{cases} \tag{4.3}$$

and the weight of **c** *is*

$$wt_H(\mathbf{c}) = \sum_{j=0}^{n-1} wt(c_j) \tag{4.4}$$

Definition 7. *Hamming distance [67]*
The Hamming distance of two codewords $\mathbf{c_1}$ and $\mathbf{c_2}$ is:

$$dist(\mathbf{c_1}, \mathbf{c_2}) = wt_H(\mathbf{c_1} - \mathbf{c_2}) \tag{4.5}$$

Definition 8. *Minimal Hamming distance [67]*
The minimal distance of a code is the minimal distance between two different
codewords $c_1, c_2 \in C$:

$$d = \min_{\substack{c_1, c_2 \in C \\ c_1 \neq c_2}} \{dist(c_1, c_2)\} \tag{4.6}$$

4.2.3 Bose-Chaudhuri-Hocquenghem (BCH) codes

BCH codes are cyclic ECC that are capable to correct sporadic errors. In this
sub-section, we consider binary BCH codes. We use the standard notation
(n, k, d) to denote a block code $\mathcal{B}$ of length n, dimension k, and minimum
Hamming distance d where $d \geq 2t + 1$ and t is the error correction capability.
The theoretical length of the binary BCH code $\mathcal{B}$ is $n = 2^m - 1$. A BCH code
can be defined by the cyclotomic cosets $\mathcal{K}_i$ [67].

Definition 9. *Cyclotomic Coset*
Let $n = p^m - 1$. *The cyclotomic coset* $\mathcal{K}_i$ *is:*

$$\mathcal{K}_i := \{i \cdot p^j \bmod n, j = 0, 1, \ldots, m - 1\} \tag{4.7}$$

where i denotes the coset numbering.

Definition 10. *BCH generator polynomial [67]*
Let $\mathcal{K}_i$ *be the cyclotomic coset of* $n = 2^m - 1$, α *the primitive element of*
$GF(2^m)$ *and* $\mathcal{M}$ *the union of arbitrary cyclotomic cosets* $(\mathcal{M} = \mathcal{K}_1 \cup \mathcal{K}_2 \cup \ldots)$.
A primitive BCH code has the length $n = 2^m - 1$ *and the generator polynomial*
$g(x)$:

$$g(x) = \Pi_{i \in \mathcal{M}}(x - \alpha^{-i}), \mathcal{M} = \bigcup_{i=1}^{d-1} \mathcal{K}_i. \tag{4.8}$$

where d is the planned minimum distance. In order to encode the information
polynomial $i(x)$, systematic BCH encoding is performed by multiplying $i(x)$
with coefficients in GF(2) and the generator polynomial $g(x)$, so the BCH codeword $b(x) = i(x) \cdot g(x)$.

The BCH codes are nested linear codes. Let $\mathcal{B}_1$ and $\mathcal{B}_2$ be two nested codes where nested means that $\mathcal{B}_2$ is a subset of $\mathcal{B}_1$. In order to clarify the nested property of the linear code, we consider the following example.

Definition 11. *Minimal polynomials [67]*
Let $\mathcal{K}_i$ be the cyclotomic coset to $n = 2^m - 1$ and α the primitive element of $GF(2^m)$, so the minimal polynomial $m_i(x)$ is

$$m_i(x) = \Pi_{j \in \mathcal{K}_i}(x - \alpha^j) \tag{4.9}$$

Example 3. *We construct a double error correction BCH code of length $n = 15 = 2^4 - 1$ [67]. It can correct up to $t = 2$ where t is the error-correcting capability, therefore the minimum distance d must be: $d \geq 5$. The cyclotomic cosets are $\mathcal{K}_1 = \mathcal{K}_2 = \mathcal{K}_4 = \{1, 2, 4, 8\}$ and $\mathcal{K}_3 = \{3, 6, 9, 12\}$ where the union $\mathcal{M} = \mathcal{K}_1 \cup \mathcal{K}_3 = \{1, 2, 3, 4, 6, 8, 9, 12\}$. The minimum distance is $d = 5$, since the numbers $1, 2, 3, 4$ are included in $\mathcal{M}$. The dimension is $k = 15 - |\mathcal{M}| = 15 - 8 = 7$. By applying Definition 11, the minimal polynomial $m_i(x)$ have coefficients in $GF(2)$, where*

$$m_1(x) = x^4 + x + 1 \tag{4.10}$$

and

$$m_3(x) = x^4 + x^3 + x^2 + x + 1. \tag{4.11}$$

The generator polynomial $g_i(x)$ are

$$
\begin{aligned}
g_1(x) &= m_1(x) \\
g_2(x) &= m_1(x) \cdot m_3(x) = x^8 + x^7 + x^6 + x^4 + 1.
\end{aligned}
\tag{4.12}
$$

Based on the nested property, $\mathcal{B}_2 \subset \mathcal{B}_1$ since $\mathcal{B}_2$ over $GF(2)$ is dividable by the minimal polynomials $m_1(x)$ and $m_3(x)$.

4.3 Coding scheme

The basic codeword format for an error-correcting code with flash memories is illustrated in figure 4.1 a). We assume coding with an algebraic error correcting code (e.g. a BCH code) of length n and error correction capability t, but the proposed coding scheme can also be used with other error-correcting codes. The encoding is typically systematic and operates on data block sizes of 512 bytes up to 4KB. In addition to the data and the parity for error correction, typically some header information is stored which contains additional parity bits for error detection.

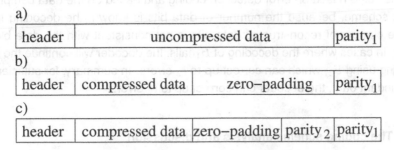

Figure 4.1: Basic codeword format.

For the applications in storage systems, the number of code bits n is fixed and cannot be adapted to the redundancy of the data. The basic idea of the proposed coding scheme is to use the redundancy of the data in order to improve the reliability, i.e. reducing probability of a decoding error by reducing the number of ones n_1 in the codeword. In order to reduce n_1 we compress the redundant data and use zero-padding as illustrated in figure 4.1 b). Furthermore, the reliability can be improved by using more parity bits and hence a higher t as indicated in figure 4.1 c). However, increasing t also increases the decoding complexity. Moreover, the error correction capability should be known for decoding the error-correcting code.

In the following, we propose a coding scheme that uses two different codes, where the decoder can resolve which code was used. For the proposed coding scheme, we use two nested codes $\mathcal{B}_1$ and $\mathcal{B}_2$ of length n and dimensions k_1 and k_2. The code $\mathcal{B}_2$ has the smaller dimension $k_2 < k_1$ and higher error correction capability $t_2 > t_1$. If the data can be compressed such that the number of compressed bits is less than or equal to k_2, we use $\mathcal{B}_2$ to encode the compressed data, otherwise we encode the data (compressed or uncompressed) using $\mathcal{B}_1$. Hence, we require an additional information bit in the header which indicates whether the data was compressed.

Because $B_2 \subset B_1$, we can also use the decoder for B_1 to decode data encoded with B_2 up to the error correction capability t_1. Thus, if the actual number of errors is less than or equal to t_1 we can successfully decode. If the actual number of errors is greater than t_1, we assume that the decoder for B_1 fails. The failure can often be detected using algebraic decoding. Moreover, a failure can be detected based on error detection coding and based on the data compression scheme, because the number of data bits is known, the decoding fails if the number of reconstructed data bits is not consistent with the data block size. In cases where the decoding of B_1 fails, the decoder will continue the decoding using B_2, which can correct up to t_2 errors. In summary, for sufficiently redundant data, the decoder can correct up to t_2 errors.

4.4 The coding scheme analysis and discussion

In this section, we present an analysis of the error probability of the proposed coding scheme for the BAC. Based on these results, we present some numerical results for an TLC flash.

4.4.1 Error analysis

We assume transmission over a BAC as illustrated in figure 4.2. It has a probability p that an input 0 will be flipped into a 1 and a probability q for a flip from 1 to 0. Furthermore, we assume for the error probabilities $q > p$.

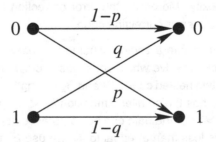

Figure 4.2: Binary asymmetric channel with error probabilities p and q.

For the BAC, the probability P_e of a decoding error depends on n_0 and $n_1 = n - n_0$, the the number of zeros and ones in a codeword. We denote the probability of i errors in the positions with zeros by $P_0(i)$. For the BAC, the number of errors for the transmitted zero bits follows a binomial distribution, i.e. the error pattern is a sequence of n_0 independent experiments, where an error occurs with probability p. We have

$$P_0(i) = \binom{n_0}{i} p^i (1-p)^{n_0-i}. \tag{4.13}$$

Similarly, we obtain

$$P_1(j) = \binom{n_1}{j} q^j (1-q)^{n_1-j} \tag{4.14}$$

for the probability of j errors in the positions with ones.

Note that the number of errors in the positions with zeros and ones are independent. Thus the probability to observe i errors in the positions with zeros and j errors in the positions with ones is $P_0(i)P_1(j)$. We consider a code with error correction capability t. For such a code, we obtain the probability of correct decoding by

$$P_c(n_0, n_1, t) = \sum_{i=0}^{t} \sum_{j=0}^{t-i} P_0(i)P_1(j) \tag{4.15}$$

and probability of a decoding error by

$$P_e(n_0, n_1, t) = 1 - P_c(n_0, n_1, t). \tag{4.16}$$

The probability $P_e(n_0, n_1, t)$ of a decoding error depends on n_0, n_1, and the error correction capability $t \in \{t_1, t_2\}$. Moreover, these values depend on the data compression. If the data can be compressed such that the number of compressed bits is less than or equal to k_2, we use $\mathcal{B}_2$ with t_2 to encode the compressed data, otherwise we encode the data using $\mathcal{B}_1$ with error correction capability $t_1 < t_2$. Hence, we define the average error probability as the expected value

$$P_e = \mathbb{E}\{P_e(n_0, n_1, t)\} \tag{4.17}$$

where the average is taken over the ensemble of all possible data blocks. In the following, we present results for empirical data. For the data model we use the

Calgary as well as the Canterbury corpus. The values of the error probabilities p and q are based on empirical data presented in [8].

4.4.2 Numerical results

Note that the error probability of a flash memory increases with the number of P/E cycles. The numbers of P/E cycles determines the life time of a flash memory, i.e. the life time is the maximum number of P/E cycles that can be executed while maintaining a sufficient low error probability. Hence, we calculate the error probability for different numbers of P/E cycles.

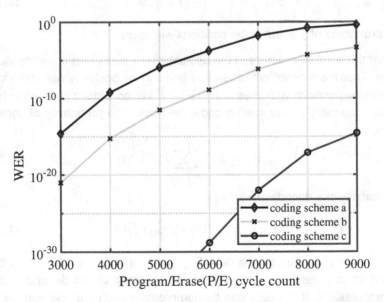

Figure 4.3: Numerical results for an TLC flash, where a, b, and c denote the coding scheme according to Figure 4.1.

The data is segmented into blocks of 1 KB, where each block is compressed and encoded independently. For ECC, we consider a BCH code $\mathcal{B}(n, k, d)$ with error-correcting capability $t_1 = 40$ if uncompressed data is encoded. This code has dimension $k_1 = 8192$ and code length is $n = 8752$. For the compressed data, we achieve a compression gain of at least 93 bytes for each data block. Hence, we can double the correcting capability and use $t_2 = 80$ with $k_2 = 7632$ (954 bytes) for compressed data. The remaining bits are filled with zero-padding as described in section 4.3. From this data processing we obtain the actual number of zeros and ones for each data block. Finally, the error probability for each block is calculated according to equation (4.17) and averaged over all data blocks.

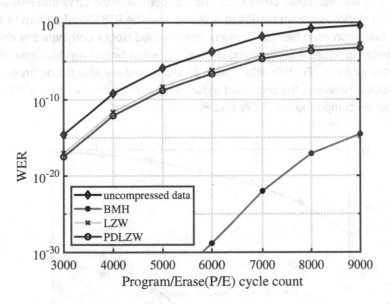

Figure 4.4: Word error rates with different data compression algorithms for the Calgary corpus.

The numerical results are presented in figure 4.3, where a, b, and c denote the coding scheme according to figure 4.1. From these results we observe that compression and zero-padding (curve b) improves the life time of the flash by more than 1000 P/E cycles compared with ECC with uncompressed data (curve a). The higher error-correcting capability (curve c) improves the life time by 4000 to 5000 P/E cycles. For this analysis, we assume perfect error detection after decoding $\mathcal{B}_1$. Hence, the word error rates (WER) are too optimistic. The actual residual error rate depends on the error detection capability of the coding scheme. Nevertheless, the error detection capability should not affect the gain in terms of P/E cycles.

Figure 4.4 depicts results for different data compression algorithm for the Calgary corpus. All results with data compression are based on the coding scheme that uses additional redundancy for error correction (coding scheme c). However, with the Calgary corpus there are blocks that might not be sufficiently redundant to add additional parity bits. This happens with the LZW and PDLZW algorithm. LZW algorithm results in 4 blocks and the PDLZW algorithm in 12 blocks uncompressed blocks. These uncompressed blocks dominate the error probability. In figure 4.5, we compared all schemes based on data from the Canterbury corpus. For this data model, all algorithms are able to compress all data blocks. However, the proposed algorithm improves the life time by 500 to 1000 cycles comparing with LZW and PDLZW schemes.

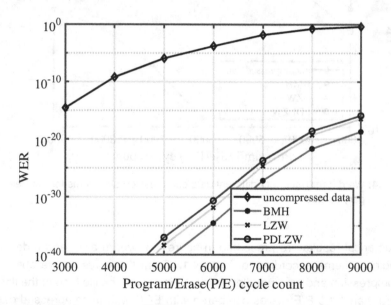

Figure 4.5: Word error rates with different data compression algorithms for the Canterbury corpus.

Finally, we consider the effect of the proposed coding scheme on data retention. Data retention is the ability to retain the programmed state. Flash data retention degrades over time and with high temperature. A charge loss induces a shift of the threshold voltage as discussed in chapter 2. The measurements presented in figure 2.6 in section 2.2 indicate that the voltage shift results in an asymmetric error probability. Figure. 4.6 presents numerical results for the empirical data presented in figure 2.6, where $0V$ represents the original read reference voltage after the first programming. The results presented in [91, 92] demonstrate that the voltage shift at room temperature increases at least linear with time. Hence, the difference in voltage shift indicates a proportional gain in data retention time. For instance, scheme b achieves a word error probability of 10^{-5} at $-0.058V$ and scheme a at $-0.038V$. Hence, scheme b achieves a gain of 50% compared with the value of coding scheme a. Scheme c achieves the same word error probability for a shift of $-0.9V$, i.e. a gain of approximately 225% compared with the value of coding scheme a. This result indicates at least a doubling of the flash data retention time with the proposed coding scheme c compared with scheme a.

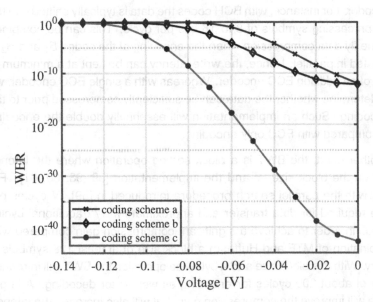

Figure 4.6: Word error rates versus voltage shift.

4.4.3 Encoding and decoding latency

The proposed source and channel coding scheme will introduce extra latency for data writing and reading. Typically, the latency for the program process is not an issue for flash applications, but the read latency affects the random access capability of the storage system. The latency depends on the data compression algorithm as well as on the implementation.

For instance, the PDLZW algorithm as well as the combination of MH can be encoded and decoded on the fly introducing only a few cycles of latency both for encoding and decoding. For the PDLZW the number of cycles depends on the number of parallel code books and varies from 4 to 8 cycles for encoding and decoding [42]. Similarly, the encoding and decoding of the Huffman code requires data buffering which introduces about 10 cycles latency [43]. Hence, the read latency with these two compression schemes is almost negligible.

The write latency depends on the combination of the data compression with the ECC encoder. For instance, with BCH codes the data is typically shifted into the encoder processing symbols of 8 to 16 bits per cycle. This can be combined with on the fly data compression when the encoding of the codes $\mathcal{B}_1$ and $\mathcal{B}_2$ is implemented in parallel. Hence, the write latency can be kept at a minimum at the price of an addition ECC encoder. Whereas with a single ECC encoder, we have to determine whether the data can be sufficiently compressed prior to the ECC encoding. Such an implementation will essentially double the encoding latency compared with ECC only encoding.

On the other hand, the BWT is a block sorting operation where the latency depends on the block size N and the implementation [59, 93, 60, 61]. For instance, with the parallel search procedure introduced in [59], N cycles per block are required for data transfer and approximately $N/2$ additional cycles for sorting. In order to achieve a significant compression gain compared with the combination of MTF and Huffman a block size of at least 128 symbols is necessary. With this block size and a bytewise operation the BWT will introduce a latency of about 200 cycles for encoding as well as for decoding. A larger block size will improve the compression gain, but will also increase the latency. The latency can be reduced with a larger word size for the BWT operation at the cost of additional logic.

4.5 Summary

In this chapter, we have presented a data compression algorithm for short data blocks which is based on BMH scheme. This compression gain is used to improve the reliability of flash memories. With the Calgary as well as the Canterbury corpus and a block size of 1 KB, the compression gain is sufficient to improve the error correction capability by 40 errors per block for each block in the corpus. For blocks with a higher compression gain, we use zero-padding which also reduces the error probability, due to the asymmetric channel. The proposed coding scheme results in a significant improvement of the life time of MLC and TLC flash memories. With redundant data, the proposed combined coding scheme results in a significant improvement of the P/E cycling endurance and the data retention time of flash memories.

The proposed combined ECC and data compression scheme can be implemented with moderate hardware requirements. For instance, the combined MTF and Huffman coding was implemented in [43] on a Xilinix Vertix-7 690T FPGA. The encoder and decoder require each about 2650 LUT and 2200 flip-flops achieving a throughput of 480 Mbit/s. On the other hand, an implementation of the BCH codec requires a total of 30600 LUT and 16300 flip-flops for the same FPGA [72]. It should be noted that the complexity of the ECC part is not increased, because the BCH codes in [72] supports multiple code rates.

The data compression improves the reliability of flash memories, by reducing WA or improving the error correction capability of the ECC unit. It is difficult to distinguish which schemes outperform the others. Both schemes are suitable for memory, where the number of errors under P/E cycles effect are clearly reduced. For flash memory, it is important to consider a number of constraints: applicable compression algorithm for short block size, compression performance, hardware complexity, high data rates, latency, and applicable strong ECC.

5 Product codes and generalized concatenated codes for flash memories

Data reliability and integrity are important requirements for storage systems and is ensured by ECC. Traditionally, a binary symmetric channel (BSC) is used as channel model for flash memories and BCH codes are used for error correction [94, 72]. Recently, different concatenated coding schemes were proposed that have low error correcting capabilities. For instance, the PC, HPC, and GCC, these codes are well suited for error correction in flash memories which require a high reliability [95, 96, 97, 98, 99] [78]. Next section, we outline a motivation of HPC and GCC that are applicable for flash memories.

In section 5.2, we investigate PC and HPC where the PC is defined as a two-dimensional array of linear block codes. A 2D PC is presented as a set of matrices where each row and column is a codeword in one constituent code [100]. We propose extended BCH codes for HPC based on an iterative anchor decoding [101, 102]. Anchor decoding aims to avoid most miscorretions across component codes. This decoder achieves a reasonable performance compared with a genie decoder that avoids all miscorrections [103]. Moreover, we present simulation performance for PC and HPC that have low error-correcting capabilities.

For concatenated coding schemes based on BCH component codes, the hardware implementations for BCH codes are based on the Berlekamp Massey algorithm (BMA). However, for single, double, and triple error-correcting BCH codes, Peterson's algorithm can be more efficient than the BMA. The known hardware architectures of Peterson's algorithm require Galois field inversion. This inversion dominates the hardware complexity and limits the decoding speed. In section 5.3, we propose an inversion-less version of Peterson's algorithm. Moreover, a decoding architecture is presented that is faster than decoders that employ inversion or the fully parallel BMA at a comparable circuit size.

© The Editor(s) (if applicable) and The Author(s), under exclusive license to
Springer Fachmedien Wiesbaden GmbH, part of Springer Nature 2020
M. Rajab, *Channel and Source Coding for Non-Volatile Flash Memories*,
Schriftenreihe der Institute für Systemdynamik (ISD) und optische
Systeme (IOS), https://doi.org/10.1007/978-3-658-28982-9_5

In section 5.4, we propose a construction for high-rate GCCs. The GCCs are constructed from inner binary BCH codes and outer RS codes. In order to enable high-rate codes, we propose extended BCH codes, where we apply SPC codes in the first level of the GCC. The simulation results of the GCC based on hard decision are presented. Finally, we summarize this chapter in section 5.5. Parts of this chapter are published in [78, 104].

5.1 Motivation

The error rate of the flash memory gradually degrades due to many error effects until the flash reaches the EOL. NAND flash memories aim to achieve a very strong error-correcting capability [105]. Such coding systems require high code rates, very high throughput with hard-input decoding, and low residual error rates. These requirements can be meet by GCCs, PCs , HPCs, or staircase codes.

Recently, PC and HPC are also considered for flash memories in [106, 107], where the HPC outperforms the PC in both waterfall and error floor regions with iterative decoding. PCs suffer from high error-floor and this may not be applicable for flash memories. However, the HPC seems attractive for flash memory, which has a low residual error rate. Many studies are focused on achieving a good trade-off between the error-floor and waterfall regions for flash memory. For instance, the proposed codes in [108, 109], take structural feature from HPC, where they provide a systematic trade-off which achieves a target WER at higher error probability. These schemes are mainly constructed based on BCH codes. For example, PC constructions based in BCH codes were proposed in [110, 111, 112, 113]. Hardware architectures for such codes were proposed for instance in [114, 115, 73, 116, 117]. Similarly, implementations for fast decoding of staircase codes requires fast BCH decoding [118, 119].

Due to the required code rates, BCH codes that can only correct single, double, or triple errors are used. The decoding of the concatenated codes typically requires multiple rounds of BCH decoding. Hence, the achievable throughput depends strongly on the speed of the BCH decoder. Moreover, BCH codes that correct only two or three errors are used in random access memory (RAM) applications [120, 121, 122], that require high data throughput and a very low decoding latency. In this chapter, we propose an inversion-less version of Peterson's algorithm for triple error-correcting BCH codes. The proposed algorithm is more efficient than the decoders employing Galois field inversion [123, 115]. Moreover, the proposed inversion-less Peterson's algorithm provides more flexibility regarding the hardware implementation and enables pipelining to speed up the decoding. A decoding architecture for such a pipelined architecture is presented.

Concatenated codes are PC that combine codes with different characteristics in one matrix. In [124], Forney shows his definition on code concatenation. Figure 5.1 shows that the outer code considers the inner code as super code. There are also different constrictions of the concatenated code, as described by Blokh and Zyablov in [125]. Both constructions consider the inner and the outer codes as a single code, which is called GCC.

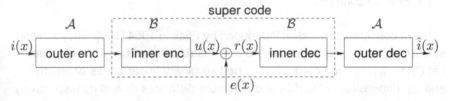

Figure 5.1: Code concatenation by Forney

Recently, for NAND flash memories concatenated codes were proposed that are constructed from long BCH codes [110, 111]. These codes can achieve low residual error rates, but require very long codes and hence a long decoding latency, which might not be acceptable for all applications of flash memories. For some applications, WER less than 10^{-16} are required [74]. For instance, the SSD requires WER of less than 10^{-15} for the client applications and less than 10^{-16} for enterprise solutions [126]. We demonstrate that GCC can achieve such low residual error rates for the hard/soft-input decoding.

Flash memory provides a spare area that is used to store the redundancy required for the ECC. This spare area determines the code rate of the error correction code. The code constructions presented in [95, 116] are limited to codes with overall code rate less than 0.90 which is not applicable in flash memories that provide only a small spare area. In this chapter, we construct the

GCC based on extended BCH codes. We extend the multi-level code construction by using SPC to enable high-rate codes and low complexity. For instance, the GCC have a low complexity compared with long BCH codes and are well suited for fast hardware decoding architectures [116, 117]. Moreover, the GCC can be easily adapted to soft-input decoding.

5.2 Product code (PC) and half-product code (HPC)

Nowadays, PCs are introduced in many application, especially for high-speed optical communication and storage systems. In this section, we provide an overview of the PCs and HPC based on the binary linear code.

5.2.1 Product code (PC)

PCs are concatenated codes, introduced by Elias in 1954 [100]. They are constructed by arranging the codeword in a matrix form with rows and columns. Let $C_1(n_1, k_1, d_1)$ and $C_2(n_2, k_2, d_2)$ be two binary linear codes of lengths n_1 and n_2, dimensions k_1 and k_2 and minimum distances d_1 and d_2 respectively. The PC $\mathcal{P} = C_1 \otimes C_2$ comprises all matrices whose rows are in C_1 and columns in C_2. The code $\mathcal{P}(C_1, C_2)$ is an (N, K, D) linear code with length $N = n_1 \cdot n_2$, dimension $K = k_1 \cdot k_2$, and minimum distance $D = d_1 \cdot d_2$, as illustrated in figure 5.2. If the component codes of the PC, given by $C_1(n_1, k_1)$ and $C_2(n_2, k_2)$, have generator matrices G_1 and G_2 respectively [102]. Subsequently, the encoding of the information matrix $U_{k_1 \times k_2}$ into a codeword $c \in \mathcal{P}$ is given by

$$c = G_1^T \cdot U \cdot G_2. \tag{5.1}$$

There are many decoders proposed for PC, although the cascaded decoder proposed in [102] was the most efficient for hardware implementation. The cascaded decoder is able to correct many error patterns beyond their correction capability [102]. The decoding is processed in an iterative procedure. An iterative row-column decoder operates on the rows and the columns successively. This process is cascaded until the decoder corrects all errors in the matrix (see figure 5.3) . The cascade decoder has so-called stopping-set patterns, which are the error patterns that lead to uncorrectable code and no further decoding.

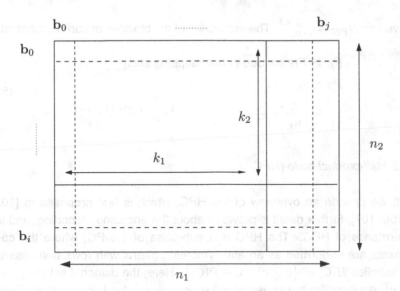

Figure 5.2: Structure of PC

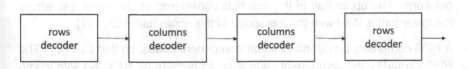

Figure 5.3: Cascaded decoding

For PC, t is the number of correctable errors for the component code. Let t_1 and t_2 be the error correction capabilities of C_1 and C_2, respectively. The stopping-set pattern happens when any error pattern with at least $t_2 + 1$ errors in each of at least $t_1 + 1$ at different columns. Thus, the minimum stopping set is $S_{PC} = (t + 1)^2$ [106]. The total number of possible minimum stopping sets is given as $N_{PC} = \binom{n}{t+1}^2$. The probability of the block error can be satisfied to

$$P_B \geq P(\text{error due to min stopping sets})$$

$$P_B \approx N_{PC} \cdot p^{SPC}$$

$$\approx \frac{1}{(n_1+1)!} \frac{1}{(n_2+1)!} n_1^{t_1+1} n_2^{t_2+1} p^{(t_1+1)(t_2+1)}. \tag{5.2}$$

5.2.2 Half-product code (HPC)

Next, we provide an overview of the HPC, which is first proposed in [103]. In [106, 107], further detail is provided about the encoding, decoding, and the performance of HPC. The HPC is a sub-class of the PC, where the code elements are performed as an anti-symmetric matrix with rows and columns that satisfies $\mathcal{H}(\mathcal{C}) = \{\mathbf{c} - \mathbf{c}^T : \mathbf{c} \in \mathcal{P}(\mathcal{C})\}$. Here, the diagonal set to $\mathbf{c}_{i,i} = 0$ and $\mathbf{c}^T$ denotes the transpose of $\mathbf{c}$, i.e. $\mathbf{c}_{i,j} = \mathbf{c}_{j,i}$ for $1 < i, j \leq n$. These conditions imply zeros on the diagonal and the symmetry of the code elements. The diagonal elements are set to zero, because they are weakly protected positions. The upper half of the matrix is considered as the message, where the lower half is filled with the transpose of the upper half [106, 107].

A HPC comprises two identical linear component codes, so that $\mathcal{C}_1 = \mathcal{C}_2$. The HPC formed by the component code $\mathcal{C}_1(n, k)$ is given by $\mathcal{H}(N, K)$ with length $N = \frac{n(n-1)}{2}$ and dimension $K = \frac{k(k-1)}{2}$. Hence, $\mathcal{H}(\mathcal{C})$ is defined by puncturing sub-code of $\mathcal{P}(\mathcal{C}_1, \mathcal{C}_2)$. The encoding of HPC is similar to the encoding of PC by forming a anti-symmetrical $k \times k$ array with zero diagonal and encodes the corresponding PC.

In order to calculate the minimum distance D_{min} of the $\mathcal{H}(\mathcal{C})$, let $supp(\mathbf{c}) \triangleq \{i \in [n] \,|\, \mathbf{c}_i \neq 0\}$ denote to support set of codeword vector $\mathbf{c}$. The second generalized Hamming weight of an (n, k, d) code satisfies $d_2 \geq \lceil 3d/2 \rceil$. Thus, the minimum distance of $\mathcal{H}$ is given by

$$D_{min} \geq \begin{cases} \frac{3d^2}{4} & \text{if } d \text{ is even} \\ \frac{(3d+1)d}{4} & \text{if } d \bmod 4 = 1 \\ \frac{(3d+1)d+2}{4} & \text{if } d \bmod 4 = 3 \end{cases} \tag{5.3}$$

Let $\mathcal{V}$ denote the anti-symmetric sub-code of $\mathcal{P}(\mathcal{C}, \mathcal{C})$. For $\mathbf{c}_1, \mathbf{c}_2 \in \mathcal{C}$, all non-zero codeword arrays of the form $\mathbf{c} = \mathbf{c}_1^T \mathbf{c}_2 \notin \mathcal{V}$. This is because $diag(\mathbf{c}) = 0$ for $\mathbf{c} \in \mathcal{V}$ and thus, $[\mathbf{c}_1]_i [\mathbf{c}_2]_i = 0$ for all i. Thus, $supp(\mathbf{c}_1) \cap supp(\mathbf{c}_2) = \varnothing$, and $\mathbf{c}_{i,j} = [\mathbf{c}_1]_i [\mathbf{c}_2]_j \neq 0$ which implies $\mathbf{c}_{j,i} = [\mathbf{c}_1]_j [\mathbf{c}_2]_i = 0$. Thus, $\mathbf{c}^T \neq -\mathbf{c}$ and $\mathbf{c} \notin \mathcal{V}$, where all non-zero rows are scalar multiples of the same codeword. The non-zero codeword in $\mathcal{V}$ must contain at least two distance non-zero rows, where the minimum number of non-zero columns is lower bounded by d_2. Likewise, each column must have at least d non-zero elements and this implies that $d_{min}(\mathcal{V}) \geq d_2 d \geq \lceil 3d/2 \rceil d$. Due to symmetry, $d_{min}(\mathcal{V})$ must be even, it can be improved to $d_{min}(\mathcal{V}) \geq 2 \lceil \lceil 3d/2 \rceil d/2 \rceil$. Since, $\mathcal{H}(\mathcal{C})$ is formed by puncturing the codeword arrays in $\mathcal{V}$, i.e. $D_{min} = d_{min}(\mathcal{V})/2$. Thus, the minimum distance of $\mathcal{H}(\mathcal{C})$ is $D_{min} \geq \lceil \lceil 3d/2 \rceil d/2 \rceil \approx 3d^2/4$ [107].

Example 4. *Given HPC $\mathcal{H}(\mathcal{C})$ with the component code $\mathcal{C}_1(n, k, d)$, where $n = 8$, $k = 4$, and $d = 4$. The codeword length of $\mathcal{H}(\mathcal{C})$ is $N = \frac{n(n-1)}{2} = 28$ and the information length is $K = \frac{k(k-1)}{2} = 6$. The minimum distance of $\mathcal{H}(\mathcal{C})$ is $D_{min} = 3d^2/4 = 12$. This codeword of the $\mathcal{H}(\mathcal{C})$ corresponds to the following matrix $\mathbf{c}$,*

$$
\mathbf{c} = \begin{bmatrix}
0 & c_{1,2} & c_{1,3} & c_{1,4} & c_{1,5} & c_{1,6} & c_{1,7} & c_{1,8} \\
c_{1,2} & 0 & c_{2,3} & c_{2,4} & c_{2,5} & c_{2,6} & c_{2,7} & c_{2,8} \\
c_{1,3} & c_{2,3} & 0 & c_{3,4} & c_{3,5} & c_{3,6} & c_{3,7} & c_{3,8} \\
c_{1,4} & c_{2,4} & c_{3,4} & 0 & c_{4,5} & c_{4,6} & c_{4,7} & c_{4,8} \\
c_{1,5} & c_{2,5} & c_{3,5} & c_{4,5} & 0 & c_{5,6} & c_{5,7} & c_{5,8} \\
c_{1,6} & c_{2,6} & c_{3,6} & c_{4,6} & c_{5,6} & 0 & c_{6,7} & c_{6,8} \\
c_{1,7} & c_{2,7} & c_{3,7} & c_{4,7} & c_{5,7} & c_{6,7} & 0 & c_{7,8} \\
c_{1,8} & c_{2,8} & c_{3,8} & c_{4,8} & c_{5,8} & c_{6,8} & c_{7,8} & 0
\end{bmatrix}. \quad (5.4)
$$

The upper triangular part of $\mathbf{c}$ determines the HPC codeword, where the $diag(\mathbf{c}) = 0$.

As it can be observed, the HPCs have larger minimum distance that PCs [106]. The minimum stopping set of the HPC is given as $S_{HPC} = \frac{(t+1)(t+2)}{2}$, where t is the error correction capability of the component code. The total number of possible minimum stopping sets is given by $N_{HPC} = \binom{n}{t+2}$. Thus, the probability of block error P_B satisfies

$$P_B \geq P(\text{error due to min stopping sets})$$
$$P_B \approx N_{HPC} \cdot p^{S_{HPC}}$$
$$\approx \frac{1}{(t+1)!} n^{t_1+1} p^{(t^2+3t+2)/2}. \tag{5.5}$$

On the other hand, the decoding of HPC is considered the codeword as full codeword form and applying a cascade decoder. Essentially, the performance of the HPC decoding can be significantly improved by applying the anchor decoding [101].

In [101], they proposed the anchor decoder algorithm in order to improve the performance of the iterative decoding of product codes over the BSC. Note that the undetected errors in the component codes called miscorrections which affect the decoding performance. The anchor decoder exploits conflicts between component code due to miscorrection where the two component codes may disagree on the value of a particular bit. Thus, the decision is made based on the number of conflicts after each component decoding whether the decoding result is indeed reliable. This designates based on status information for each component code. The number of conflicts can be reduced by keeping track of the conflict locations and avoiding the bit flips of the most reliable component codes as anchors. Therefore, it is not allowed for further additional correction from other component code, if this leads to a conflict. Since some anchors can be miscorrected. Therefore, anchor decoder allows to backtrack the decoding decisions of anchors. Backtracking is performed when a certain number of component codes report conflicts with the specific anchor.

As a results, the anchor decoder archives a significant performance improvement in the waterfall and error-floor especially for small error correction capability $t \in 2,3$. On the other hand, the complexity of the anchor decoder is increased. For further information, the anchor decoder and the pseudocode can be found in [101] for iterative bounded distance decoding (BDD). In the next section we apply the anchor decoder for PCs and HPCs in order to avoid the genie decoder which prevents miscorrection.

5.2.3 Simulation results for BCH-PC and BCH-HPC

We propose simulation results for HPC based on the extended BCH component code. These results are compared with the PC results and with genie decoder from [107]. Note that PC and HPC decoding are performed with iterative hard-decision decoding. The performance of the component decoding significantly degrades from undetected errors, i.e. miscorrections. Therefore, we propose a simulation results for HPC based on anchor decoding [101], where we apply single- and double-extended BCH codes. The single-extended BCH is obtained through an additional parity bit by adding (modulo 2) all coded bits of the BCH code. In contrast, the double-extended BCH has two additional parity bits, where the parity bits perform checks on odd and even bit positions separately. For PCs and HPCs, the channel is considered to be a channel with symbol error probability ϵ. A Monte Carlo simulation is performed to estimate the BER and WER for different ϵ.

In order to achieve a fair comparison between a BCH-PC and BCH-HPC, we use the same code-rate and block length as in [107]. Let BCH code be primitive brainy code, i.e. $n = 2^m - 1$. Subsequently, BCH-HPC is $\mathcal{H}(N, K, D)$, where $N = n(n-1)/2$, and $K = k(k-1)/2$. Let the component shortened primitive binary BCH-PC be $\mathcal{P}(N', K', D')$ with $N' = (n')^2$ and $K' = (k')^2$. Note that the minimum stopping set size of HPC increases by $\approx 0.12t'$, where t' is the error correction capability of the PC [107].

First, we compare the PC parameters as proposed in [107], where the PC based on a shortened BCH(170, 154, 5) with $t' = 2$. The BCH-PC has $\mathcal{P}(28900, 23716, 25)$ with code rate $R' = 0.82$ and a minimum-size stopping set of 9. Similarly, we use a single-extended BCH(171, 154, 6) and double-extended BCH(171, 153, 6) as component code. The single- and double-extended BCH-PC have $\mathcal{P}(29241, 23716, 36)$ and $\mathcal{P}(29241, 23409, 36)$, with $R' = 0.81$ and $R' = 0.80$, respectively. The code rate of the extended component code reduces but still can be comparable to the ordinary component.

Figure 5.4 presents the BER and the WER performance for BCH-PC. The idealized decoding of the PC provides a significant performance due to the genie decoder prevents all miscorrections. Since there is no genie decoder in reality, the results are optimal and not realistic. With realistic decoders the decoding error rate increases. In figure 5.4, the ordinary BCH-PC has the worst

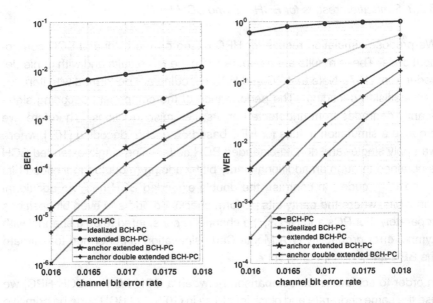

Figure 5.4: Performance comparison between ordinary BCH-PC, single- and double-extended BCH-PC with/without anchor decoder for $t = 2$. The left figure depicts the performance based on the BER while the right figure depicts the performance based on the WER.

performance, which is caused by miscorrections. Next, we apply a single-extended BCH-PC which provides better performance compared with ordinary BCH-PC. The gap between ordinary/single-extended BCH-PC and genie decoder remains large due to miscorrection cases. In order to reduce this gap, we apply an anchor decoding based on the single- and double-extended BCH-PC. The double-extended BCH-PC has clearly significant performance compared with ordinary and single-extended BCH-PC. However, the error performance become close to the genie decoder, where the anchor decoder reduces the miscorrection cases. This encourages to apply the anchor decoder to the HPC, where the error performance can be improved.

Second, we consider HPC that has equivalent parameters to the PC. Therefore, the HPC chosen based on the BCH$(255, 231, 7)$ code and has parameters

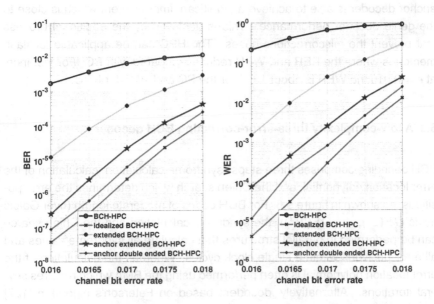

Figure 5.5: Performance comparison between ordinary BCH-HPC, single- and double-extended BCH-HPCwith/without anchor decoder for $t = 3$. The left figure depicts the performance based on the BER while the right figure depicts the performance based on the WER.

$\mathcal{H}(32385, 26565, 37)$ with $t = 3$ and $R = 0.82$. Furthermore, the single- and double-extended BCH parameters are $BCH(256, 231, 8)$ and $BCH(256, 230, 8)$, respectively. Thus, the components for single- and double-extended codes are $\mathcal{H}(32640, 26565, 48)$ and $\mathcal{H}(32640, 26335, 48)$ with $R = 0.81$ and $R = 0.80$, respectively. Figure 5.5 shows the simulated performance of the HPC, where it can be seen clearly that the HPC outperforms the PC. This is likely due to the change of the error correction capability t from 2 to 3 which improves HPC performance at the same code rate.

In figure 5.5, the ordinary BCH-HPC has the worst error probability performance. By applying the single- and double-extended BCH-HPC, the error probability performance clearly improves. In order to avoid the genie decoder, we propose the extended component code based on an anchor decoding. The anchor decoder is able to achieve a significant improvement which is close to the genie decoder performance. This is achieved by the additional process that prevent the miscorrections cases. The HPC can be applicable for flash memories where the BER and WER reduce compared with PC. For instance, at $\epsilon = 0.016$ the WER is about 10^{-3} for the PC and 10^{-5} for HPC.

5.3 A low-complexity three-error-correcting BCH decoder

BCH decoding comprises three steps: syndrome calculation, calculation of the error location polynomial, and the Chien search which determines the error positions, as shown in figure 5.6. For BCH codes of moderate length (over Galois fields $GF(2^6), \ldots, GF(2^{12})$), the syndrome calculation and the Chien search can be performed in parallel structures that calculate all syndrome values and all error positions within a single clock cycles. Whereas the calculation of the error location polynomial is often performed using the BMA which requires several iterations. Alternatively, decoders based on Peterson's algorithm [127] were proposed in [123, 128, 114, 115]. Such decoders can be more efficient than the BMA for BCH codes with small error correcting capabilities, i.e. single, double, and triple error-correcting codes.

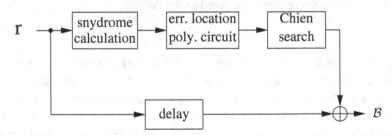

Figure 5.6: Structure of a BCH decoder.

5.3.1 Peterson's algorithm

In this sub-section, we briefly revise Peterson's algorithm and introduce the notations. The received vector is $r(x) = b(x) + e(x)$, where $b(x) = b_0 + b_1 x +$

$\ldots + b_{n-1}x^{n-1}$ is a codeword of length n and $e(x) = e_0 + e_1 x + \ldots + e_{n-1}x^{n-1}$ is the error vector. $S_1, S_2, \ldots, S_{2t-1}$ denote the syndrome values which are defined as

$$S_i = r(\alpha^i) = e(\alpha^i), \tag{5.6}$$

where α is the primitive element of the Galois field $GF(2^m)$. For binary BCH codes, the following relation holds

$$S_{2i} = S_i^2. \tag{5.7}$$

Let ν be the actual number of errors and t is the error-correcting capability of the BCH code. The coefficients of the error location polynomial $\sigma(x) = \sigma_0 + \sigma_1 x + \ldots + \sigma_t x^\nu$ satisfy a set of equations called Newton's identities. In matrix form these equations are

$$\mathbf{A}_\nu \mathbf{\Delta}_\nu = \mathbf{S}_\nu. \tag{5.8}$$

With $\sigma_0 = 1$, the $(i \times i)$ matrix

$$\mathbf{A}_i = \begin{pmatrix} 1 & 0 & 0 & 0 & \ldots & 0 \\ S_2 & S_1 & 1 & 0 & \ldots & 0 \\ S_4 & S_3 & S_2 & S_1 & \ldots & 0 \\ \vdots & \vdots & \vdots & \vdots & \ldots & \vdots \\ S_{2i} & S_{2i-1} & S_{2i-2} & \ldots & \end{pmatrix}, \tag{5.9}$$

the vector of coefficients

$$\mathbf{\Delta}_i = \begin{pmatrix} \sigma_1 \\ \sigma_2 \\ \vdots \\ \sigma_i \end{pmatrix}, \tag{5.10}$$

and the syndrome vector

$$
\mathbf{S}_i = \begin{pmatrix} -S_1 \\ -S_2 \\ \vdots \\ -S_{2i+1} \end{pmatrix}.
\tag{5.11}
$$

Note that the matrix $\mathbf{A}_i$ is singular for $i > \nu$. Hence, Peterson's algorithm first calculates the number of errors ν. Starting with $i = t$ the determinant $D_i = \det(\mathbf{A}_i)$ is calculated. If $D_i = 0$ then the algorithm reduces the matrix $\mathbf{A}_i$ until $D_i = \det(\mathbf{A}_i) \neq 0$ holds and equation (5.8) can be solved.

Finally, the Chien search determines the error positions by searching for the roots of the error location polynomlal. The calculation of $\sigma(\alpha^i)$ for $i = 0, \ldots, n-1$ can be conducted in parallel using simple logic operations [116, 120].

5.3.2 Calculating the error location polynomial for low error capability

For single, double, and triple errors the following direct solutions of the Newton's identities follow [129]

$$
\sigma(x) = 1 + S_1 x \quad \text{for } \nu = 1
\tag{5.12}
$$

$$
\sigma(x) = 1 + S_1 x + \frac{S_3 + S_1^3}{S_1} x^2 \quad \text{for } \nu = 2
\tag{5.13}
$$

$$
\sigma(x) = 1 + S_1 x + \frac{S_1^2 S_3 + S_5}{S_3 + S_1^3} x^2 +
$$
$$
\left(S_1^3 + S_3 + S_1 \frac{S_1^2 S_3 + S_5}{S_3 + S_1^3} \right) x^3 \quad \text{for } \nu = 3
\tag{5.14}
$$

These solutions are used in [123, 130, 115] for decoding BCH codes. The main difference between [123] and [115] is the implementation of the Galois field inversion in equation (5.14). For instance, in [115] a parallel hardware implementation is proposed. This architecture requires only four Galois field multipliers, but additionally a Galois field inversion is required. The complexity and the throughput of this architecture are determined by the inversion. For the Galois field $GF(2^{10})$ the size of the inversion is about twice the size of a multiplier and the length of the critical path is four times longer than that of a multiplier. In [123] the inversion is implemented using a look-up table, which is only feasible for small GF, because the table size is of order $O(m2^m)$.

In the following, we propose an algorithm for triple errors that omits the Galois field inversion similar to the approach in [130] that considers double errors. Omitting inversion reduces the hardware complexity and speeds up the calculation. First, we consider the case for single and double errors. Not that the roots of the error location polynomial do not change, if we multiply all coefficients with a non-zero factor. For instance, multiplying the right hand side of equation (5.13) with $S_1 \neq 0$, we obtain an equivalent solution

$$\sigma(x) = S_1 + S_1^2 x + D_2 x^2 \qquad (5.15)$$

for $\nu = 2$ with the determinant

$$D_2 = S_3 + S_1^3. \qquad (5.16)$$

Note that for $\nu = 1$ and $\nu = 2$, S_1 is non-zero. For a single error in position e_1 we have $S_1 = \alpha^a \neq 0$. Similarly, for two errors in positions e_1 and e_2, we have $S_1 = \alpha^{e_1} + \alpha^{e_2} \neq 0$, because $\alpha^{e_1} \neq \alpha^{e_2}$. Equation (5.15) is also a solution for $\nu = 1$, because $D_1 = S_1 \neq 0$ and $D_2 = 0$ holds for a single error.

Next, we consider the case $\nu \geq 2$. For $\nu = 2$ and $\nu = 3$, we have $D_2 \neq 0$ [123]. To observe this, first consider $\nu = 2$, we have $S_1 = \alpha^{e_1} + \alpha^{e_2}$ and $S_3 = \alpha^{3e_1} + \alpha^{3e_2}$. Hence,

$$\begin{aligned} S_1^3 + S_3 &= (\alpha^{e_1} + \alpha^{e_2})^3 + \alpha^{3e_1} + \alpha^{3e_2} \\ &= \alpha^{e_1 + 2e_2} + \alpha^{2e_1 + e_2} \neq 0 \text{ for } e_1 \neq e_2. \end{aligned} \qquad (5.17)$$

Similarly, for $\nu = 3$ we have $S_1 = \alpha^{e_1} + \alpha^{e_2} + \alpha^{e_3}$ and $S_3 = \alpha^{3e_1} + \alpha^{3e_2} + \alpha^{3e_3}$. Consequently,

$$\begin{aligned} S_1^3 + S_3 &= (\alpha^{e_1} + \alpha^{e_2} + \alpha^{e_3})^3 + \alpha^{3e_1} + \alpha^{3e_2} + \alpha^{3e_3} \\ &= \alpha^{e_1 + 2e_2} + \alpha^{e_1 + 2e_3} + \alpha^{e_2 + 2e_3} + \\ &\quad \alpha^{e_2 + 2e_1} + \alpha^{e_3 + 2e_1} + \alpha^{e_3 + 2e_2}. \end{aligned} \qquad (5.18)$$

The last term is the determinant of the following matrix

$$\begin{pmatrix} 1 & \alpha^{e_1} & \alpha^{2e_1} \\ 1 & \alpha^{e_2} & \alpha^{2e_2} \\ 1 & \alpha^{e_3} & \alpha^{2e_3} \end{pmatrix}. \tag{5.19}$$

This matrix has full rank, because the columns are linearly independent. Hence, $D_2 \neq 0$ holds for $\nu = 3$. Now, multiplying the right hand side of equation (5.14) by D_2, we obtain an equivalent solution for $\nu = 3$ as

$$\sigma(x) = D_2 + S_1 D_2 x + \delta_2 x^2 + D_3 x^3 \tag{5.20}$$

with

$$\delta_2 = S_1^2 S_3 + S_5 \tag{5.21}$$

and the determinant

$$D_3 = S_1(S_2 S_3 + S_1 S_4) + S_3^2 + S_1 S_5. \tag{5.22}$$

Using Equations (5.6), (5.16), and (5.21) we obtain

$$\begin{aligned} D_3 &= S_1(S_1^2 S_3 + S_5) + S_1^6 + S_3^2 \\ &= S_1 \delta_2 + D_2^2. \end{aligned} \tag{5.23}$$

The decoding procedure is summarized in algorithm 1. This algorithm can easily be adapted to the decoding of single and double error-correcting codes, e.g. setting $D_3 = 0$ for double error-correcting BCH codes. This is important for the iterative anchor decoding of the PC and HPC, where the error correction capability is low.

Algorithm 1 Inversion-less Peterson algorithm
---

    calculate D_2, δ_2, D_3
    if $D_3 == 0$ **then**
        return $\sigma(x) = S_1 + S_1^2 x + D_2 x^2$
    else
        return $\sigma(x) = D_2 + S_1 D_2 x + \delta_2 x^2 + D_3 x^3$

5.3.3 Hardware architecture

In this sub-section, we present a hardware architecture for the proposed decoding algorithm and compare its speed (critical path length) and the array consumption with other algorithms. Not that the critical path length and the circuit size is dominated by the Galois field multipliers and Galois field inversion. The size of a bit-parallel multiplier grows with order $O(m^2)$ and the critical path with $O(m)$. The Galois field inversion is often implemented using Fermat's little theorem, which requires only a single multiplier and a squaring operation, but $m - 1$ clock cycles [131]. Hence, the total number of basic logic operations per inversion is of order $O(m^3)$. On the other hand, the addition and squaring operations are of order $O(m)$ with a critical path length $O(1)$. Consequently, these two operations are neglected in the following discussion.

Algorithm 1 can be implemented performing all operations in parallel. Such an implementation requires four multipliers and has a critical path length of two multipliers. It is more efficient than the implementation proposed in [115]. The architecture in [115] uses three multipliers and one inversion, where the logic for the inversion is about twice the size of a multiplier (for $GF(2^{10})$) and the critical path length of the inversion is equivalent to four multiplications. The total critical path in [115] has a length that is equivalent to six multiplications. Hence, at a smaller size the proposed algorithm has a significantly shorter critical path.

Moreover, the proposed algorithm enables pipelined architectures that can speed up the decoding. Whereas the Galois field inversion is an atomic operation which limits the efficiency of pipelining. Figure 5.7 present such a pipeline (without control logic). The pipeline requires four multipliers and additional registers (three registers of width m bits) to store intermediate results. The pipeline reduces the critical path length to a single multiplication. Hence, the pipelined architecture doubles the throughput compared with the structure without pipeline. Note that the fully parallel BMA also has a critical path length of a single multiplication. However, the parallel BMA requires $2t$ multipliers, at least $2t$ registers, and t iterations [71, 72]. Hence, the proposed architecture is smaller and about three times as fast as the parallel BMA. In order to verify the above size considerations, the proposed decoding algorithm has been implemented on a FPGA in Verilog. Table 5.1 contains results for the Xilinx Virtex-7

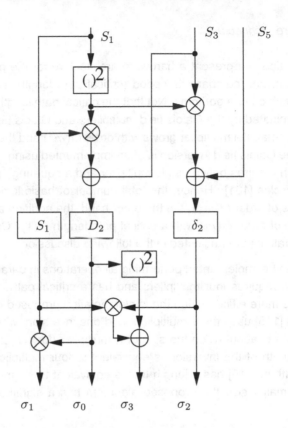

Figure 5.7: Hardware architecture of the decoder pipeline.

FPGA. The fundamental building blocks of an FPGA are flip-flops and the LUTs. The size of the logic is represented by the number of LUT. Table 5.1 represents data for $m = 8$ and $m = 12$. As can be seen, both decoders require $3m$ flip-flops for the registers. The size of the decoder for $m = 12$ is dominated by the four multipliers which require about 90% of the logic. The speed of the decoders is determined by the achievable clock frequency f_{clk}. The circuit for $m = 8$ achieves a clock frequency $f_{clk} = 500$MHz, i.e. a throughput of $500 \cdot 10^6$ BCH codewords per second, because one codeword is processed per clock cycle. The latency is two clock cycles, i.e. 4ns. Moreover, the decoder for $m = 8$ is about 1.5 times faster than the decoder for $m = 12$, which

Table 5.1: Results for the FPGA implementation for the proposed algorithm.

Module	number of LUT	number of flip-flops	throughput codewords per second
proposed decoder $GF(2^8)$	145	24	$500 \cdot 10^6$
multiplier $GF(2^{12})$	71	–	$333 \cdot 10^6$
proposed decoder $GF(2^{12})$	318	36	$333 \cdot 10^6$
parallel BMA $GF(2^{12})$	426	84	$111 \cdot 10^6$

confirms that the critical path length is of order $O(m)$. Similarly, the ratio of 2.2 for the logic size for $m = 12$ and $m = 8$ agrees well with the estimate $O(m^2)$ for the circuit size. The results for the parallel BMA are estimates based on the required number of multipliers and registers. The actual size will be higher. The BMA requires three clock cycles per BCH codeword. Hence, the achievable throughput is $111 \cdot 10^6$ BCH codewords per second at a clock frequency of $f_{clk} = 333$MHz.

5.4 Generalized concatenated codes (GCC)

In this section, we consider GCCs for error correction in flash memories that require high-rate codes. The GCCs are constructed from inner nested binary BCH codes and RS codes. For the inner codes we propose single- and double-extended BCH codes. In order to enable high-rate GCC, we apply SPC codes in the first level of the GCC. We present a decoding method for the GCCs and an analysis of the decoding error probability.

5.4.1 Reed-Solomon codes

First, we outline the basic aspects of the RS codes, as proposed by Reed and Solomon in 1960. RS codes are applied in many telecommunication applications. RS codes are popular in ECCs, where they are suitable for burst errors [89]. We apply the RS as an outer component for GCC.

Definition 12. *RS code [89]*
RS code is BCH code p^m-ary of length $n = p^m - 1$, and the minimum distance

of an $\mathcal{A}(n, k)$ RS is $d = n - k + 1$. The minimal polynomial in GF(p^m) of an element β in GF(p^m) is simply $(x - \beta)$. The RS codeword is $a(x) = i(x) \cdot g(x)$, and the generator for a RS code is defined as,

$$g(x) = \prod_{j=n-k}^{n-1} (x - \alpha^{-j}) \qquad (5.24)$$

where α is a primitive element and the code has $k = p^m - 1 - 2t$ information symbols. The parity-check polynomial $h(x)$ is defined as ,

$$h(x) = \prod_{j=0}^{k-1} (x - \alpha^{-j}). \qquad (5.25)$$

Algebraically, we can decode the received vector $r(x) = a(x) + e(x)$ using syndrome calculating $S(x)$. If $S(x) = 0$ then $r(x)$ is a valid codeword, otherwise, we have to determine the error locations by using BMA as for BCH codes. Additionally, we have to calculate the error value with so-called Forney algorithm. This can be derived by calculating the error value polynomial that solves the so-called Berlekamp-Forney key equation. At the end, we use Chien search to determine the roots of the error locator polynomial.

5.4.2 GCC construction

In this sub-section, we describe the GCC construction and its parameters. A detailed discussion can be found in [132] or [67]. We propose GCCs with inner extended BCH codes similar to the construction in [78]. However, we consider GCC with short inner binary BCH codes, where we use the SPC codes in the first level. Later, we consider all component codes have the same alphabet similar to the generalized error-locating (GEL) codes proposed in [97]. One reason for the longer inner codes is the decoding complexity, given that most GCCs use outer RS codes with a larger alphabet size than the inner BCH codes. With this construction the decoder complexity is dominated by the logic for the RS decoder [116, 117]. Using stronger inner codes reduces the decoder complexity for the outer codes. Later on, we will show that the longer inner component codes also improve the decoding performance compared with the codes in [78].

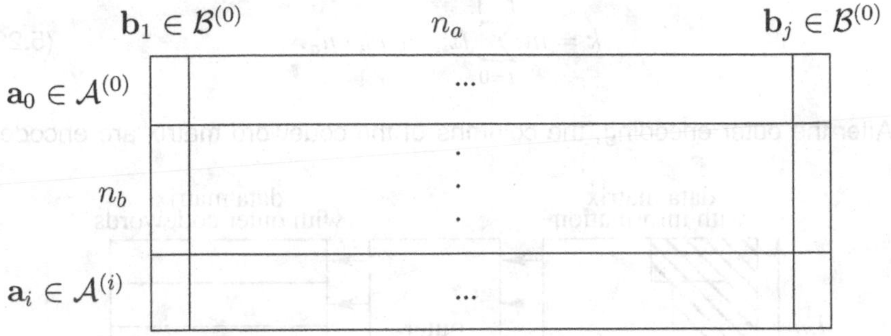

Figure 5.8: Codeword matrix.

The GCC is a multi-level code which has two component codes for each dimension. Figure 5.8 illustrates the GCC codeword matrix, which is an $n_b \times n_a$ matrix. The parameters n_a and n_b denote the lengths of the outer codes $\mathcal{A}^{(l)}$ and inner codes $\mathcal{B}^{(l)}$, respectively. All component codes are constructed over $GF(2^m)$.

5.4.3 GCC encoding

The encoding starts with the outer codes. The rows of the codeword matrix are protected by L RS codes of length n_a, i.e. L denotes the number of levels. m elements of each column represent one symbol from the $GF(2^m)$. Hence, m rows form a codeword of an outer code $\mathcal{A}^{(l)}, l = 0 \ldots, L-1$. Note that the code rate of the outer codes increases from level to level. The outer codes protect Lm rows of the matrix. The code dimensions are $k_b^{(0)} = Lm, k_b^{(1)} = (L-1)m, \ldots, k_b^{(L-1)} = m$. The remaining $n_b - Lm$ rows are used for the redundancy of the inner codes or for information bits without outer encoding. The shaded area in figure 5.9 illustrates the redundancy of the component codes that are filled by outer and inner encoding. Let k_u denote the number of rows without outer encoding. Subsequently, the dimension of the GCC is

$$k = m \sum_{l=0}^{L-1} k_{a,j} + k_u \cdot n_a. \tag{5.26}$$

After the outer encoding, the columns of the codeword matrix are encoded

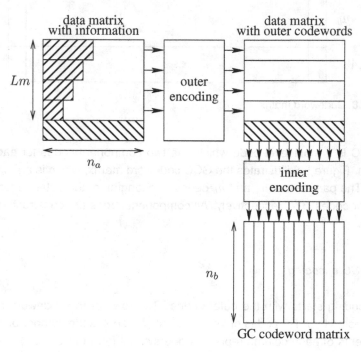

Figure 5.9: GC encoding scheme.

with binary inner codes of length n_b. In this work, we consider extended BCH codes as inner codes. Each column of the codeword matrix is the sum of $L + 1$ codewords of nested extended BCH codes. This GCC encoding is not systematic. In order to read the information the codeword has to be reimaged.

$$\mathcal{B}^{(L)} \subset \mathcal{B}^{(L-1)} \subset \ldots \subset \mathcal{B}^{(0)} \tag{5.27}$$

Hence, a higher level code is a sub-code of its predecessor, where the higher levels have higher error-correcting capabilities, i.e. $t_{b,L-1} \geq t_{b,L-2} \geq \ldots \geq t_{b,0}$, where $t_{b,l}$ is the error-correcting capability of level l. The codeword of the j-th column is the sum of L codewords.

$$\mathbf{b}_j = \sum_{l=0}^{L} \mathbf{b}_j^{(l)}. \tag{5.28}$$

These codewords $\mathbf{b}_j^{(l)}$ are formed by encoding the symbols $a_{j,l}$ with the corresponding sub-code $\mathcal{B}^{(l)}$, where $a_{j,l}$ is the j-th symbol (m bits) of the outer code $\mathcal{A}^{(l)}$. For this encoding $(L-l-1)m$ zero bits are prefixed onto the symbol $a_{j,l}$. Note that the j-th column $\mathbf{b}_j$ is a codeword of $\mathcal{B}^{(0)}$, due to the linearity of the nested codes. Figure 5.10 illustrates the resulting column codewords for three levels, where $\mathbf{p}_{j,i}$ denotes the parity of the codeword $\mathbf{b}_j^{(i)}$.

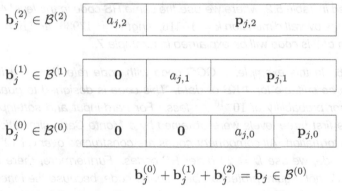

$$\mathbf{b}_j^{(0)} + \mathbf{b}_j^{(1)} + \mathbf{b}_j^{(2)} = \mathbf{b}_j \in \mathcal{B}^{(0)}$$

Figure 5.10: Nested BCH codeword

In the following examples, we consider the GCC notion as $\mathcal{C}(n, k, d)$, to denote a code $\mathcal{C}$ of length n, dimension k, and the minimum distance d over the GF(2^m). The length of $\mathcal{C}$ is $n = n_a \cdot n_b$ with $k = m \sum_{l=0}^{L-1} k_{a,l}$ and $d = min_l\{d_{a,l}, d_{b,l}\}$ where $d_{a,l}$ and $d_{b,l}$ are the minimum Hamming distance at level l of the RS and BCH code respectively.

Example 5. *The GCC is designed for 2 KB information blocks, i.e. a code which can be used to store 2 KB of data plus 4 bytes of meta information. For this GCC, we use $L = 13$ levels with binary BCH codes over GF(2^7) and outer*

Table 5.2: Parameters of the code from example 5

level l	$k_{b,l}$	$d_{b,l}$	$k_{a,l}$	$d_{a,l}$
0	117	2	84	69
1	108	4	130	23
2	99	6	136	17
3	90	8	142	11
4-12	81	12	148	5

RS codes over $GF(2^9)$. *Hence, the dimension of the inner codes is reduced by* $m = 9$ *bits with each level. In the first level, we apply SPC codes. Hence, all inner codes are extended BCH codes of length* $n_b = 13 \cdot 9 + 1 = 118$ *and the outer RS codes of length* $n_a = 152$. *The parameters of the codes are summarized in Table 5.2, where we use the same RS code in the level 4 to 12. The code has overall dimension* $k = 16416$, *length* $n = 17936$, *and* $R = 0.915$. *The design of this code will be explained in Example 7.*

Example 6. *In this example, a GCC code with code rate* $R = 0.9$ *is constructed to be suitable for 4 KB of data. This code is designed to guarantee a word error probability of* 10^{-16} *or less. For hard-input and soft-input decoding, the first three levels were obtained by a Monte Carlo simulation with reliability information. All component codes are constructed over* $GF(2^8)$. *For this GCC code, we use* $L = 20$ *outer RS codes. Furthermore, there are 10 levels that do not require protection be an outer code, because the inner code guarantees the required word error probability. The outer RS codes have a length of* $n_a = 143$. *The inner codes are binary extended BCH codes of length* $n_b = 256$. *Table 5.3 summarizes the parameters of the GCC code for hard-input and soft-input decoding. For hard-input decoding, we use the same RS code from level 8 to level 10 and from 11 to 19. For the soft-input decoding, we use the same RS code from 3 to 5, where the transition from hard-input to soft-input decoding happened. Overall, the code has dimension* $k = 32953$, *length* $n = n_a \cdot n_b = 36608$, *and rate* $R = \frac{k}{n} = 0.90$ *for both hard/soft-input decoding. The design of this code will be explained in Example 8 and Example 9.*

Table 5.3: Parameters of the code from example 6.

(a) Parameters of the hard-input code

level l	$k_{b,l}$	$d_{b,l}$	$k_{a,l}$	$d_{a,l}$
0	247	4	121	23
1	239	6	61	83
2	231	8	29	115
3	223	10	23	121
4	215	12	15	129
5	207	14	11	133
6	199	16	9	135
7	191	18	7	137
8-10	207	14	5	139
11-19	199	16	3	141

(b) Parameters of the soft-input code

level l	$k_{b,l}$	$d_{b,l}$	$k_{a,l}$	$d_{a,l}$
0	247	4	111	33
1	239	6	55	89
2	231	8	29	115
3-5	223	10	17	127
6	215	12	11	133
7	207	14	9	135
8-9	199	16	7	137
10-12	191	18	5	139
13-19	183	20	3	141

5.4.4 GCC decoding

We consider GCC for both cases, i.e. hard- and soft-decision decoding. In this chapter, we only explain the hard-input decoding. The soft-input decoding is considered in the following chapter. In practice, the soft information is only used for blocks where the decoding without reliability information fails.

Figure 5.11 illustrates the GCC decoding steps, where the decoder processes level by level starting with $l = 0$. Let l be the index of the current level and r be the faulty received $n_b \times n_a$ data matrix. First, the columns are decoded with respect to $\mathcal{B}^{(i)}$. For the inner decoding, we exploit the structure of the extended BCH codes. Let $d_{b,l}$ denote the minimum Hamming distance of the l-th inner code. The extended BCH codes have an even minimum Hamming. The error correction capability $t_{b,l}$ of the inner code of level l is

$$t_{b,l} = \frac{d_{b,l}}{2} - 1. \tag{5.29}$$

Hence, the inner code can correct an error pattern with up to $t_{b,l}$ errors with algebraic decoding [67]. Furthermore, any error pattern with $t_{b,l} + 1$ can be detected. In this case a decoding failure is declared for the inner code, where

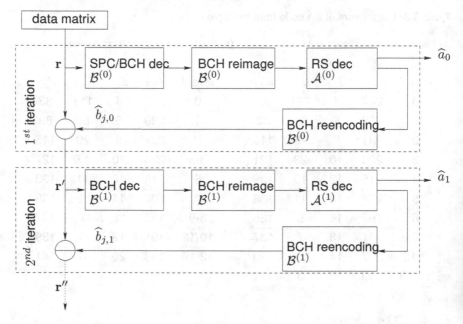

Figure 5.11: GCC decoding scheme

the decoding failure of the inner code is regarded as erased symbols of the outer RS code. Hence, error and erasure decoding is used in all levels of the RS code.

After inner decoding, the information bits of the inner codes have to be inferred (re-image) to retrieve the code symbols $a_{j,i}$ of $\mathcal{A}^{(i)}$ where j is the column index. If all symbols of the code $\mathcal{A}^{(i)}$ are inferred the RS code can be decoded. At this point a partial decoding result $\hat{a}_i$ is available. Finally this result has to be re-encoded using $\mathcal{B}^{(i)}$. The estimated codewords of the inner code $\mathcal{B}^{(i)}$ are subtracted from the codeword matrix before the next level can be decoded. This is repeated for each l of the codeword $\mathcal{C}$. Once all l codes are subtracted form r and the correction was successful, the remaining GCC data matrix only contains the errors that occurred in the channel. Since all $\hat{a}_i$ are available, the complete codeword is generated by combining all $\hat{a}_i$ together.

5.4.5 Probability of decoding errors and a code design rule

In this sub-section, we present an analysis of the probability of a decoding error for the GCC decoder with inner extended BCH codes. The error probability is required for the industrial applications. For flash memories, The maximum uncorrectable WER must be guaranteed. This can be determined by using the binomial distribution for given a specific code parameter, and the target WER.

First, we consider the probability of a decoding error with this multi-stage decoding algorithm with hard-input decoding. We assume a BSC, i.e. errors occur statistically independent with error probability ϵ. Let $P_{b,l}$ be the error probability for the decoding of the inner code $\mathcal{B}^{(l)}$. Furthermore, let $P_{e,l}$ be the corresponding probability of a decoder failure.

A decoding error with inner decoding occurs only if the number of errors in the j-th column is greater than or equal to $t_{b,l}+2$, whereas an decoding failure may occur for $t_{b,l}+1$ or more errors. Hence, we can bound the error and erasure probabilities for the inner single-extended BCH codes by

$$P_{b,l} \leq \sum_{j=t_{b,l}+2}^{n_b} \binom{n_b}{j} \epsilon^j (1-\epsilon)^{n_b-j} \tag{5.30}$$

and

$$P_{e,l} \leq \sum_{j=t_{b,l}+1}^{n_b} \binom{n_b}{j} \epsilon^j (1-\epsilon)^{n_b-j}. \tag{5.31}$$

In order to bound the error probability for double-extended BCH codes, we can detect exactly half of the error pattern with wight $t+2$, by

$$P_{b,l} \leq \frac{1}{2} \cdot \left(\sum_{j=t_{b,l}+2}^{n_b} \binom{n_b}{j} \epsilon^j (1-\epsilon)^{n_b-j} \right). \tag{5.32}$$

The additional erasure probability are included in the bound of equation (5.31).

Similarly, we bound the probability of an outer decoding error. Let $T_l = n_a - k_{a,l}$ be the number of redundancy symbols for the outer RS code $\mathcal{A}^{(l)}$ at the l-th level. The probability $P_{a,l}$ of a decoding error with error and erasure decoding at the l-th level can be computed as follows [133]:

$$P_{a,l} \leq \sum_{q=0}^{T_l} \sum_{t=\lfloor \frac{T_l-q}{2} \rfloor+1}^{n_a-q} P_q \binom{n_a-q}{t} P_{b,l}^t (1-P_{b,l})^{n_a-q-t}$$

$$+ \sum_{q=T_l+1}^{n_a} P_q \tag{5.33}$$

where P_q is the probability of q erasures

$$P_q \leq \binom{n_a}{q} P_{e,l}^q (1-P_{e,l})^{n_a-q}. \tag{5.34}$$

Using the union bound, we can estimate the WER P_e for the GCC, i.e. the likelihood of the event that at least one level is in error

$$P_e \leq \min(\sum_{l=0}^{L-1} P_{a,l}, 1). \tag{5.35}$$

Based on inequality equation (5.35), we propose a design rule for GCC similar to the equal error probability rule proposed in [134] for multi-level coding. Accordingly, we choose the code rates of the outer RS codes such that the word error probability bounds of the different levels are equal for a given channel error probability ϵ or a given signal to noise ratio (SNR) value. This rule leads to a code design that minimizes the decoding error probability with the proposed multi-stage decoding procedure. The code from example 5 and example 6 were designed according to this rule to guarantee a word error probability 10^{-16} or less.

In the next chapter, we propose a soft-input decoding method. With soft-input decoding, the error and erasure probabilities $P_{b,l}$ and $P_{e,l}$ of the inner codes can be determined using Monte Carlo simulation. Hence, we can also use equation (5.33) and equation (5.35) to estimate the performance of the GCC with soft-input decoding.

5.4.6 Simulation results for GCC

First, we compare the error correction performance of the GCC with the performance of long BCH codes with algebraic decoding. As a performance measure, we use the code rate that is required to guarantee an overall WER less than 10^{-16} for a given channel error probability.

Example 7. *Consider the code from Example 5. This code has a code rate $R = 0.915$ and was designed to guarantee $P_e \leq 10^{-16}$ according to Equation (5.35) for $\epsilon \leq 0.003$.*

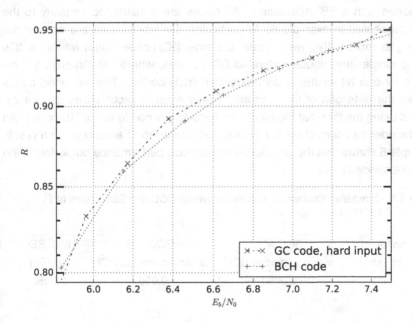

Figure 5.12: Comparison of GCC and BCH codes.

Figure 5.12 presents the code rate versus SNR for GCC as well as for long BCH codes. The code rate was chosen such that the WER is less than 10^{-16} for a given channel error probability. All codes are constructed similar to the code presented in example 5. In particular, the inner codes are chosen according to Table 5.2. Whereas the error-correcting capability of the outer codes are adapted to obtain the highest possible code rate for a channel error probability. Note that in this example, the overall code rate of the GCC is at most $R = 0.99$, due to the choice of the inner code. Note that the codes presented in [95, 116] have a code rate less than or equal to 0.90.

Example 8. *Consider the code from Example 6. This code has a code rate $R = 0.90$ and was designed according to the equal error probability rule for $\epsilon \leq 0.0043$ and $\epsilon \leq 0.0046$ for single- and double-extended BCH code, respectively, which guarantees to achieve $P_e \leq 10^{-16}$.*

In order to improve the GCC performance, the code is designed for flash memories with 4 KB information. All codes are constructed similarly to the code presented in example 6. In particular, the inner codes are chosen according to Table 5.3(a), we consider the inner BCH code length with $n_a = 256$ for the single- and double-extended BCH codes, where the dimension k reduces by one bit for the double-extended BCH codes. The extended codes enable the detection of the erasure. These erasure probabilities can be exploited using the fact that the algebraic decoding procedures for RS codes. An RS decoder can correct up to t errors, corrects up to $2t$ erasures. As a result, example 6 shows that the double-extended code performance outperforms the single-extended code.

Table 5.4: Comparison numerical results for overall GCC and SBC codes at $P_e \leq 10^{-10}$.

channel error bit rate	GCC with single-extended BCH	GCC with double-extended BCH	HPC, SBC in [109]
ϵ	0.0060	0.0067	0.0050

In order to compare the GCC with HPC schemes based on hard-decision decoding, we compare the performances of the GCC with the proposed symmetric block-wise concatenation into HPC codes, so-called symmetric BC-BCH code (SBC) in [109]. This code has better performance in terms of error floor compared with ordinary HPC, which take advantage of the block-wise and symmetry. In [109], the HPC schemes based on BCH or RS component suffer form higher error-floors. Table 5.4 consider code rate $R = 0.885$ and target

$P_e \leq 10^{-10}$ for a practical message length of 4KB. The GCC with the single and double-extended code outperforms SBC code about 0.2 and 0.5 in dB. Meanwhile, the SBC codes with 2KB and 4KB may suffer from error floors as observed in [109]. This code may not be applicable for flash memory at the EOL cases, where the error probability $\epsilon \geq 0.0050$.

5.5 Summary

In this chapter, we have proposed a construction of the PC and the HPC. The latter offers a significant improvement compared with the PC. In order to have a realistic error performance for the PC and the HPC, we applied the extended binary BCH component code, which outperforms the ordinary BCH component code. The error performance is clearly affected by miscorrections, i.e. undetected errors. In order to minimize this effect, we used the anchor decoding, which improves the performance compared with the extended versions. Furthermore, we proposed the single- and double-extended BCH codes for the HPC which the error performance outperforms the PC and gets close to genie-aided decoder that avoids all miscorrection in (cf. [107]).

In order to reduces the complexity of the BCH codes, we have proposed an algorithm to compute the error location polynomial for single, double, and triple error-correcting binary BCH codes. The proposed method is an inversion-less version of Peterson's algorithm. For triple errors the proposed algorithm is more efficient than the BMA. The pipelined decoding architecture is faster than decoders that employ inversion or the fully parallel BMA at a comparable circuit size. The new decoder can be applied for decoding the BCH component codes in concatenated codes [110, 112]. For instance, we applied the BCH decoder for PC and HPC, with error correction capability $t \in 2, 3$. The proposed algorithm reduces the complexity and speeds up the iterative decoding.

The GCCs are constructed from inner nested binary BCH codes and outer RS codes. For the inner codes, we propose extended BCH codes, where we apply SPC codes in the first level of the GCC, which enables high-rate codes. Furthermore, the single- and double-extended BCH codes enable the detection of decoding failures for the inner codes, where the outer RS codes can correct up to t errors and $2t$ erasures. We have presented a decoding method for the GCC codes and derived a bound on the decoding error probability. The proposed codes are well suited for error correction in flash memories for high reliability data storage, because very low residual error probabilities can be guaranteed. The GCC codes have a performance similar to that of BCH codes, but can be decoded faster and with a lower decoder complexity (cf. [116]). The GCCs achieved better performance compared with ordinary PC and HPC where both codes are suffer from higher error-floors which may be not applicable for flash memories [109]. Moreover, the GCCs with single- and double-extended out-perform the symmetric block-wise concatenated HPC-BCH codes proposed in [109].

With GCC that employ nested inner BCH codes [116, 135, 97] it is import-ant that the decoder supports different error-correcting capabilities, because the error-correcting capability increases from level to level. Furthermore, the proposed low-complexity BCH decoder may help to speed up soft-input de-coding algorithms for GCC that are based on Chase decoding [117] [136]. The Chase decoding procedure requires multiply BCH decoding operations for each received BCH codeword, where the calculation of the error location polynomial limits the achievable throughput. Due to the complexity of the soft-input decoding and the small performance gain for the better protected levels, the Chase decoding is limited to the first three levels in [117]. These are single, double, and triple error-correcting BCH codes. The soft-decision of the GCC decoding is presented in the following chapter.

6 A soft-input decoder of GCC for application in flash memories

In recent years, mostly hard-input algebraic decoding was used for error correction in flash memories [137, 94, 72]. ECC based on soft-input decoding can significantly improve the reliability of flash memories.

The GCC was presented in the previous chapter, where the extended BCH codes enabled an efficient hard-input decoding. The BCH codes also enable a low-complexity bit-flipping decoding. This decoder uses a fixed number of test patterns and an algebraic decoder for soft-decoding. An acceptance criterion for the final candidate codeword is proposed. This acceptance criterion can improve the decoding performance and reduce the decoding complexity. The presented simulation results show that the proposed bit-flipping decoder in combination with outer error and erasure decoding can outperform maximum likelihood (ML) decoding of the inner codes. Moreover, we provide simulation results based on the TLC flash, where we guarantee to achieve a very low residual error.

This chapter is outlined as follows. Next section, we present a motivation to explain why soft-input decoding is suited for GCCs. We describe the channel model in section 6.2. In section 6.3, the proposed bit-flipping decoding of binary block codes is presented. Afterward, the acceptance criterion for the trade-off between the error and erasure probabilities is discussed in the same section. Later, we present simulation results for the GCC with soft-input decoding. We compare the performance of GCC over AWGN channel information with ML and the Kaneko decoding method. The proposed channel estimation method is presented in section 6.4. In the same section, we provide a performance comparison between the estimation method and the perfect channel information. These are followed by a summary in section 6.5. Parts of this chapter are published in [138, 3, 136, 139, 140].

M. Rajab, *Channel and Source Coding for Non-Volatile Flash Memories*, Schriftenreihe der Institute für Systemdynamik (ISD) und optische Systeme (IOS), https://doi.org/10.1007/978-3-658-28982-9_6

6.1 Motivation

The performance of ECC can be improved if reliability information about the state of the cell is available [70]. In order to exploit the reliability information soft-input decoding algorithms are required. For instance, LDPC can provide strong error-correcting performance in NAND flash memories [79, 80, 141, 81, 142, 143]. However, LDPC codes can have high residual error rates (the error-floor) and may not be suitable for applications that require very low decoder failure probabilities. Moreover, the implementation complexity of an LDPC decoder can be high when high data throughput should be achieved. For instance, in [95] it was shown that a GCC can outer-perform a concatenated coding scheme with outer BCH code and inner regular low-density generator matrix (LDGM) code for low WER due to the error floor of the LDGM code.

In the previous chapter, the GCC with inner extended BCH codes where the SPC was applied in the first level. This construction enables high-rate GCCs. However, with soft-input decoding the inner SPC limits the coding gain [144]. In this chapter, we propose a similar GCC with inner extended BCH codes. In order to improve the decoding performance, we use longer inner component codes. In particular, we consider GCC codes with component codes over the same alphabet, similar to the GEL codes proposed in [97]. Moreover, we propose a low-complexity bit-flipping decoder to enable soft-decoding. There exist numerous soft-input decoding algorithms for binary block codes (see [67] for an overview). We propose an decoder, where the binary inner codes are decoded with a Chase decoding algorithm [145, 146, 147]. We adapt the Chase decoding for the decoding of the concatenated code. In particular, we propose a decoding procedure that declares a decoding failure if no reliable codeword is found. These failures are considered as erasures for the outer RS codes. The acceptance criterion is based on Yamamoto and Itoh's decision rule [148], which is a simplification of Forney's criterion on the trade-off between erasure and error probabilities [149].

Yamamoto and Itoh's rule requires list-of-two decoding to obtain metric values for the ML codeword and the second best codeword. Hence, Yamamoto and Itoh's decision rule is not applicable to sub-optimum decoding. In order to solve this issue, we derive an equivalent rule for the metric proposed in [146] for bit-flipping decoding. Based on an estimate of the metric value of the second best codeword, an acceptance criterion for the result of the bit-flipping is proposed. Combined with error and erasure decoding for the outer RS codes, the proposed bit-flipping decoding can significantly improve the soft-input decoding performance of GCC. It offers a performance similar to that of ML decoding of the inner codes. Finally, we compare the proposed soft-input decoding for GCC with a soft-input decoding based on sequential stack decoding of the inner codes proposed in [76].

In contrast to algebraic hard-input decoding, soft-input decoding algorithms typically require channel state information, e.g. the bit error probability or the noise variance. Essentially, the flash channel can be considered as binary AWGN channel, where the channel output is quantized using a small number of bits [110]. For instance, in [150, 151], the error characteristics of the flash channel are statistically independent and that an AWGN channel model can be assumed. As we observed from the previous chapters, the MLC and TLC require asymmetric models to characterize the errors of the flash channel [3, 43] [66, 20, 21, 8]. This error characteristic is not considered by the simple AWGN channel model. Moreover, the probabilities vary during the lifetime of a flash memory. The error probabilities and the shape of the voltage distribution depend on the number of P/E cycles and the number of reads after programming. For instance, in [152, 23] an exponential distribution is used to model the change of the voltage distribution due to electron capture/release in oxide traps that depend on the number of P/E cycles.

In this chapter, we consider a data-driven approach, i.e. we consider voltage distributions obtained by measurements. We present a method to estimate the channel state information for the different quantization levels. Our approach is histogram based, similar to the methods proposed in [153, 23], where the complete distribution of all charge levels is estimated simultaneously. In contrast to [153] and [23], we consider a setup where only statistical data for a single decision threshold is available, i.e. a page-wise read operation. Furthermore, we assume that the reading thresholds are pre-determined, e.g. in order to optimize the soft-input decoding performance, whereas in [23] an adaptive bin placement is used to optimize the channel estimation. Based on measurements, we use k-means clustering to obtain data sets that can be used to estimate the channel state.

6.2 Channel model for NAND flash memories

In this section, we provide a channel model for an approximating asymmetric error characteristic. The probability density function of the variation of threshold voltages is usually modeled by a Gaussian distribution [20, 21], (see equation (2.1)). A flash memory with reliability information can be considered as a quantized channel values. In order to obtain reliability information, the cell is read several times with small changes of the threshold voltages.

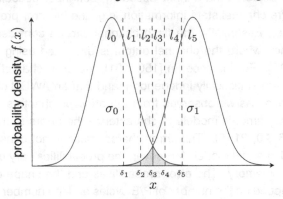

Figure 6.1: Probability density function with reading thresholds.

Figure 6.1 shows the probability distribution with five threshold voltages for the reading reference voltage, where the reading threshold voltages are denoted by δ_i with $\delta_1 < \delta_2 < \ldots < \delta_V$. We have $V + 1$ different voltage intervals l_i with $i \in \{0, \ldots, V\}$. Hence, in figure 6.1 δ_3 denotes the reference voltage R_j for the hard decision and the intervals l_2, l_3 correspond to the most unreliable input values. In this case, we can calculate the conditional probability of observing a value in the interval from level δ_i and δ_{i+1} given the charge level with mean μ as

$$w_i(\mu, \sigma) = \int_{\delta_i}^{\delta_{i+1}} f(x)dx \qquad (6.1)$$

where σ is standard deviation of the distribution. Assuming that information bits 0 and 1 are stored with equal probability, we obtain the total probability

$$q_i = \frac{1}{2}w_i(\mu_0, \sigma_0) + \frac{1}{2}w_i(\mu_1, \sigma_1) \qquad (6.2)$$

to observe a value in the interval l_i. Furthermore, we obtain the conditional error probability given that we observe a value in the interval l_i

$$p_i = \begin{cases} \frac{w_i(\mu_1, \sigma_1)}{w_i(\mu_0, \sigma_0) + w_i(\mu_1, \sigma_1)} & , i \le \frac{V}{2} \\ \frac{w_i(\mu_0, \sigma_0)}{w_i(\mu_0, \sigma_0) + w_i(\mu_1, \sigma_1)} & , i > \frac{V}{2} \end{cases} \quad (6.3)$$

Here, μ_0 and μ_1 are the mean values for the information bit 0 and 1 that is inferred with the particular reading. Likewise, σ_0 and σ_1 denote the corresponding standard deviations. The variance depends on the charge level, i.e. some charge levels are less reliable. Moreover, the error probability is not equal for zeros and ones. Note that this model is a simplification of the Gaussian mixture model presented in [153], where we assume that the probabilities p_i depend only on the two charge levels adjacent to the considered reference threshold. The reliability information is typically represented by LLR, where we have the LLR for the interval l_i

$$\iota_i = \log\left(\frac{1 - p_i}{p_i}\right). \quad (6.4)$$

In order to improve the decoding performance with quantized reliability information, let the received symbols be y_i. The quantized AWGN function is,

$$\tilde{\iota}_i = \begin{cases} \iota_1 & , y_i < \delta_1 \\ \iota_2 & , \delta_1 \le y_i < \delta_2 \\ \iota_3 & , \delta_2 \le y_i < \delta_3 \\ \iota_4 & , \delta_3 \le y_i < \delta_4 \\ \iota_5 & , \delta_4 \le y_i < \delta_5 \\ \iota_6 & , y_i \le \delta_5 \end{cases} \quad (6.5)$$

where ι_i denotes the i-th channel value and $\tilde{\iota}_i$ the corresponding quantized channel value. We optimize the values δ_i for soft-input decoding. The choice of the reading thresholds enables a trade-off between the error and erasure probabilities. Ideally, the reading thresholds would be adapted to the SNR such that the erasure probability is twice the error probability. However, with flash memories the SNR is typically not known prior the reading of the soft information. We use the thresholds $\{\delta_1, \delta_2, \delta_3, \delta_4, \delta_5\} = \{-0.40, -0.15, 0, 0.15, 0.40\}$, respectively, which where found by exhaustive search. These values provide a good compromise for the considered SNR range.

6.3 Soft-input bit-flipping decoding of GCC

In the section, we discuss soft-input decoding for the inner codes. Similar to the hard-input decoding, we aim on exploiting the error and erasure decoding of the outer codes. We investigate the soft-input decoding for quantized reliability information. In particular, we explain the bit-flipping decoding and afterward we discuss the acceptance criterion.

6.3.1 Bit-flipping decoding for the inner code

Consider a binary BCH code $\mathcal{B}$ (n, k, d), where n_b is the codeword length, k_b is the dimension, and d_b denotes the minimum Hamming distance. We assume transmission of a binary codeword $\mathbf{x} = (x_1, \ldots, x_n) \in \mathcal{B}$ over the AWGN channel. At the receiver, we obtain reliability information $(\iota_1 \ldots, \iota_n)$ from the quantized received symbols $\mathbf{y} = (y_1, \ldots, y_n)$ as represented by LLRs in equation (6.4) and the hard decision

$$\hat{y}_i = \left\{ \begin{array}{l} 1,\ \iota_i \leq 0 \\ 0,\ \iota_i > 0 \end{array} \right. \tag{6.6}$$

The vector $\hat{\mathbf{y}} = (\hat{y}_1, \ldots, \hat{y}_n)$ denotes the vector of hard decisions. Furthermore, let $(\hat{\iota}_1, \ldots, \hat{\iota}_n)$ be the ordered sequence of the LLRs such that $|\hat{\iota}_1| \leq |\hat{\iota}_2| \leq \ldots \leq |\hat{\iota}_n|$.

Bit-flipping decoding of binary block codes was pioneered by Chase [145]. Chase introduced reliability information based decoding procedures that generate a list of candidate codewords by flipping bits in the received word. The test patterns for the bit-flipping are based on the least reliable positions of the received word. For instance, the p-Chase algorithm decodes the 2^p binary n-tuples obtained by considering all possible values in the p least reliable positions [154]. The algebraic hard-input decoding is applied for each test pattern. At the end, the most likely codeword is selected from this list of candidates.

For the AWGN channel, the most likely codeword can be determined by minimizing the Euclidean distance between the received word and the codewords in the list of candidates. Alternatively, Kaneko *et al.* devised the following metric [146]

$$l(\mathbf{y}, \mathbf{x}) = \sum_{x_i \neq \hat{y}_i} |l_i| \tag{6.7}$$

where by minimizing $l(\mathbf{y}, \mathbf{x})$, the best candidate can be found. Moreover, Kaneko *et al.* introduced a stopping criterion which is a sufficient condition that the codeword with ML is found. Let $M = d_H(\hat{\mathbf{y}}, \mathbf{x})$ be the Hamming distance between the codeword $\mathbf{x}$ and the vector of hard decisions $\hat{\mathbf{y}}$. If the codeword $\mathbf{x}$ satisfies

$$l(\mathbf{y}, \mathbf{x}) \leq \sum_{i=1}^{d-M} |\hat{l}_i| \tag{6.8}$$

then there is no codeword which is more likely than $\mathbf{x}$ [146]. This criterion was used in [146] to obtain an ML decoding procedure with a variable number of decoding iterations.

For GCC, we apply the bit-flipping decoding to the inner BCH code in the GCC construction. For the inner decoder, the variable number of decoding iteration is not ideal. Hence, we use a fixed number of decoding cycles which can efficiently be implemented in a pipeline [116]. Similarly, we used the p-Chase algorithm which considers 2^p test patterns. Introducing error and erasure decoding for the outer RS code, an erasure decoding of the inner BCH decoder can even improve the decoders performance if the failures mainly happen when the ML decision is unreliable. Hence, we propose a bit-flipping procedure with a fixed number of decoding cycles and apply a decision rule to determine the final candidate whether this codeword is accepted or declared an erasure. The decision rule is discussed in the next section.

6.3.2 Decoding with erasures

Decoding algorithms with erasures provide a trade-off between erasure probability and undetected error probability after decoding. In [149], Forney derived the optimal decision rule, where a codeword $\mathbf{x}'$ should be accepted if

$$\log \frac{P(\mathbf{y}|\mathbf{x}')}{\sum_{\mathbf{x}\in\mathcal{C},\mathbf{x}\neq\mathbf{x}'} P(\mathbf{y}|\mathbf{x})} \geq T \tag{6.9}$$

where $\mathbf{x}'$ is the ML codeword, otherwise an erasure is declared. The threshold T is a trade-off parameter between error and erasure probability, where no other decision rule provides a better trade-off. However, Forney's rule is not feasible in practice. In [148], Yamamoto and Itoh simplified Forney's rule to

$$\log \frac{P(\mathbf{y}|\mathbf{x}')}{P(\mathbf{y}|\mathbf{x}'')} \geq T, \tag{6.10}$$

where $\mathbf{x}''$ denotes the second best codeword. This rule can be combined with Viterbi decoding and is motivated by the observation that the term $P(\mathbf{y}|\mathbf{x}'')$ typically dominates the sum $\sum P(\mathbf{y}|\mathbf{x})$ over all codewords $\mathbf{x} \neq \mathbf{x}'$.

Yamamoto and Itoh's rule requires list-of-two decoding and is not applicable to sub-optimum decoding, but it provides some insight into the decision problem. We first derive the following equivalent rule using Kaneko's metric

Proposition 1. *For the AWGN channel with quantized output, the rule*

$$l(\mathbf{y}, \mathbf{x}'') - l(\mathbf{y}, \mathbf{x}') \geq T. \tag{6.11}$$

is equivalent to inequality Equation (6.10).

Proof. From the definition of the metric follows

$$l(\mathbf{y}, \mathbf{x}'') - l(\mathbf{y}, \mathbf{x}') = \sum_{x_i'' \neq \hat{y}_i} |\iota_i| - \sum_{x_i' \neq \hat{y}_i} |\iota_i|. \tag{6.12}$$

The terms in both sums cancel for all values with $x_i' = x_i''$. Hence, we have

$$l(\mathbf{y}, \mathbf{x}'') - l(\mathbf{y}, \mathbf{x}') = \sum_{x_i'' \neq \hat{y}_i \wedge x_i'' \neq x_i'} |\iota_i| - \sum_{x_i' \neq \hat{y}_i \wedge x_i'' \neq x_i'} |\iota_i|. \tag{6.13}$$

Now consider Yamamoto and Itoh's rule [149]. The AWGN channel with quantized output is a discrete memoryless channel (DMC). The probability of receiving the vector $\mathbf{y}$ given the input $\mathbf{x}$ is

$$P(\mathbf{y}|\mathbf{x}) = \prod_{i=1}^{n} P(y_i|x_i). \tag{6.14}$$

Hence, we have

$$
\begin{aligned}
\log \frac{P(\mathbf{y}|\mathbf{x}')}{P(\mathbf{y}|\mathbf{x}'')} &= \sum_{i=1}^{n} \log P(y_i|x_i') - \sum_{i=1}^{n} \log P(y_i|x_i'') \\
&= \sum_{x_i' \neq x_i''} \log P(y_i|x_i') - \log P(y_i|x_i'') \tag{6.15} \\
&= \sum_{x_i' \neq x_i''} \log \frac{P(y_i|x_i')}{P(y_i|x_i'')} \tag{6.16}
\end{aligned}
$$

where Equation (6.15) holds, because the terms for $x_i' = x_i''$ in both sums cancel. For $x_i' \neq x_i''$, we have

$$\log \frac{P(y_i|x_i')}{P(y_i|x_i'')} = \begin{cases} |\iota_i|, & x_i' = \hat{y}_i \\ -|\iota_i|, & x_i' \neq \hat{y}_i \end{cases} \tag{6.17}$$

Using Equation (6.15), we obtain

$$\log \frac{P(\mathbf{y}|\mathbf{x}')}{P(\mathbf{y}|\mathbf{x}'')} = \sum_{x_i' = \hat{y}_i \wedge x_i'' \neq x_i'} |\iota_i| - \sum_{x_i' \neq \hat{y}_i \wedge x_i'' \neq x_i'} |\iota_i| \tag{6.18}$$

which is equal to Equation (6.13). □

Yamamoto and Itoh's rule cannot be applied to the bit-flipping decoding, because it requires list-of-two decoding to determine $l(\mathbf{y}, \mathbf{x}'')$. However, we can obtain an estimate of $l(\mathbf{y}, \mathbf{x}'')$. Consider for instance the case, where the hard decision $\hat{y}$ is a codeword, i.e. $\hat{y} \in \mathcal{B}$. We have $\mathbf{x}' = \hat{y}$ and $l(\mathbf{y}, \mathbf{x}') = 0$. The metric for the second best codeword satisfies $l(\mathbf{y}, \mathbf{x}'') \geq \sum_{i=1}^{d} |\hat{\iota}_i|$, because $\mathbf{x}''$ has at least Hamming distance d to the codeword $\mathbf{x}'$. Therefore, $l(\mathbf{y}, \mathbf{x}'')$ is the sum of at least d absolute values $|\iota|$. This sum can be bounded from below by the sum over the d smallest absolute values. Similarly, we have $l(\mathbf{y}, \mathbf{x}'') \geq \sum_{i=1}^{d-M} |\hat{c}_i|$ if the best codeword has Hamming distance $M = d_H(\hat{\mathbf{y}}, \mathbf{x}')$. This estimate agrees with equation (6.8), but will typically underestimate the value of $l(\mathbf{y}, \mathbf{x}'')$, because $\sum_{i=1}^{d-M} |\hat{c}_i|$ is a lower bound that holds for all possible received vectors $\mathbf{y}$. Instead of equation (6.8), we use the estimate $l(\mathbf{y}, \mathbf{x}'') \approx \sum_{i=1}^{M} |\hat{c}_i|$, where M is a parameter (independent of the distance $d_H(\hat{\mathbf{y}}, \mathbf{x}')$) that can be adjusted to the channel condition (e.g. to the expected number of errors). We propose the following decision rule

$$l(\mathbf{y}, \mathbf{x}') \leq \sum_{i=1}^{M} |\hat{c}_i| - T \qquad (6.19)$$

where the codeword $\mathbf{x}'$ is the final candidate of the bit-flipping decoding and is accepted if the inequality holds. The threshold parameter T and the parameter M enable a trade-off between the error and erasure probability. A larger value of M leads to a higher acceptance rate and hence decreases the erasure probability.

6.3.3 Simulation results for an acceptance criterion over AWGN channel

In this sub-section, we present simulation results for the symmetric AWGN channel, i.e. $\mu_0 = 1, \mu_1 = -1$ and $\sigma_0 = \sigma_1$. Note that the probability of a decoding error with error and erasure decoding at the outer codes can be computed based on the error and erasure probabilities of the inner codes. With bit-flipping decoding, these probabilities can be determined by Monte Carlo simulation.

In order to compare the performance with bit-flipping decoding to that of ML decoding, we first consider short binary BCH codes of length $n_b = 64$. Later on, we present results for GCC codes suitable for ECC in flash memories, where we use inner codes of length $n_b = 256$.

First, we consider the decoding of the binary BCH codes with the proposed acceptance criterion using different trade-off parameters M and T. Figure 6.2 shows simulation results for the BCH code $\mathcal{B}(64, 57, 4)$. For this simulation, $p = 3$ was used. It can be seen, that the ratio of the erasure and error probabilities does hardly vary with the SNR. It is determined by the trade-off para-

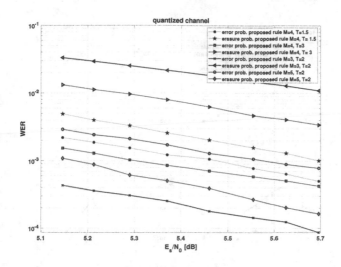

Figure 6.2: Simulation results for BCH code $\mathcal{B}\,(64, 57, 4)$ for different trade-off parameters.

meters M and T. Larger values of M and T lead to lower erasure probabilities, but higher error probabilities.

In figure 6.3 and figure 6.4, we compare the decoding performance of the proposed decoding algorithm with the performance of ML decoding and that of Kaneko's rule for a fixed number of test patterns ($p = 3$ for both bit-flipping algorithms). Figure 6.3 depicts simulation results for the BCH code $\mathcal{B}(64, 57, 4)$ for the AWGN channel without quantization. Similarly, figure 6.4 presents results for the channel with quantized output values. In order to compare the performance with outer decoding, we calculate the error probability of the RS code $\mathcal{A}(285, 260, 25)$ according to equation (5.33) in chapter 5.

Note that ML decoding always results in a valid codeword. Hence, the erasure probability is zero. On the other hand, Kaneko's rule results in a high erasure probability that dominates the error probability of the outer RS decoding with this small set of test patterns. By applying the acceptance criterion of equation (6.19), we can lower the erasure probability compared with Kaneko's rule. Using $M = 4, T = 0.3$ for the channel without quantization and $M = 4$,

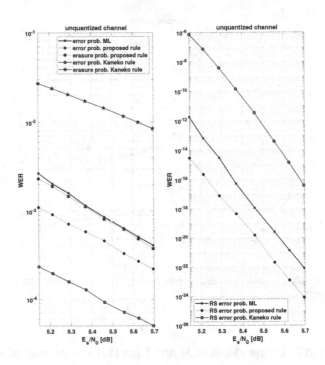

Figure 6.3: Simulation results for BCH code $\mathcal{B}(64, 57, 4)$ without quantization (the left figure depicts the performance of inner BCH decoding, the right figure demonstrates the word error probability after RS decoding)

$T = \min(|\iota_i|)$ with quantized channel result in the smallest error probability after outer decoding with the proposed rule. For the channel without quantization, the proposed acceptance criterion achieves a gain of about 0.5 dB compared with Kaneko's rule and 0.1 dB compared with ML decoding. With quantized channel values, the gain compared with Kaneko's rule is smaller, but the proposed decoding outperforms ML decoding by 0.1 dB.

Figure 6.5 provides simulation results for the second level codes (BCH code $\mathcal{B}(64, 51, 6)$ and RS code $\mathcal{A}(285, 274, 9)$) for the channel with quantization. For this simulation, $p = 3$ was used for the proposed decoding and $p = 4$ for the decoder with Kaneko's rule. As can be seen, a list with eight patterns ($p = 3$) is

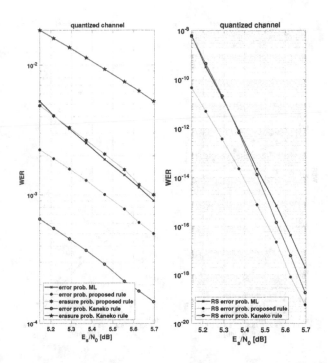

Figure 6.4: Simulation results for BCH code $\mathcal{B}(64, 57, 4)$ for the quantized channel (the left figure depicts the performance of inner BCH decoding, the right figure demonstrates the word error probability after RS decoding)

sufficient with the proposed rule to obtain a near-ML performance. Even with 16 test patterns ($p = 4$), Kaneko's rule leads to a high erasure probability that impairs the RS decoding performance. The proposed rule reduces the complexity compared with Kaneko's rule and improves the decoding performance with outer RS decoding.

Finally, we consider some results for the GCC codes. Table 6.1 shows the minimum SNR value to achieve a WER less than 10^{-16} for the quantized AWGN channel, where the code from example 6 in the previous chapter was used with the proposed decoding method for $R = 0.9$. Additionally, we provide results for the overall code rate $R = 0.93$. The GCC codes are constructed for 4 KB of

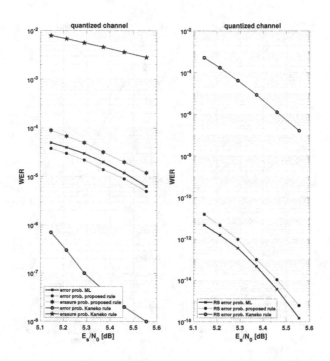

Figure 6.5: Simulation results for the second level codes for the quantized channel (the left figure depicts the word error probability with the inner BCH code $\mathcal{B}(64, 51, 6)$, the right figure the word error probability after RS decoding with the code $\mathcal{A}\ (285, 274, 9)$).

Table 6.1: Numerical results for overall GCC decoding.

code rate	hard-input BCH decoding	hard-input GCC decoding	soft-input GCC decoding [144]	proposed soft-input GCC decoding
0.90	6.2	5.4	5.3	4.9
0.93	6.7	6	5.9	5.5

data. The proposed soft-input decoding method is compared with the decoding performance of the bit-flipping decoding proposed in [144]. For hard-input decoding, we compare the performance to that of long BCH codes of the same rate.

As can be seen the new construction achieves almost the same performance with hard-input decoding as the GCC codes from [144] with soft-input decoding. Using the proposed bit-flipping decoder, an additional gain of about 0.4 dB is achieved. The proposed codes perform 1.3 dB better than the long BCH codes.

6.4 Channel estimation for NAND flash memories

In this section, we propose the channel estimation method with quantized threshold. Later, we compare the performance with perfect channel information to that with the proposed estimation method.

6.4.1 Channel estimation

The objective of the channel estimation is to determine the LLR values for the different quantization intervals. Essentially, this could be achieved by using pilot data that is stored in the flash and counting bit errors conditioned on the observation interval for the pilot data. The conditional error probabilities are typically very low. Therefore, many pilot bits would be required to estimate the corresponding LLR values. Our channel estimation approach is based on the observation that an estimate of the total probabilities q_i can be obtained by counting the number of occurrences for each quantization interval during the reading process.

In order to map the estimated $\tilde{q}_i$ values to LLR values, we consider a data-driven approach. Accordingly, we assume that a number of measurements of the actual voltage distributions of threshold values for the different charge levels are available. For such a measurement, we can calculate the total probabilities q_i, the conditional error probabilities p_i, and the ι_i values as discussed in section 6.2. Next, we use k-means clustering (cf. [155]) of the different data sets to obtain a number of representatives for each data set. The k-means clustering is applied independently for each reference threshold, because the distributions vary from threshold to threshold. Table 6.2 presents examples of measurement results for a TLC flash, where each row is a representative.

Based on this set of representatives, the channel estimation can be performed by a table look-up procedure. During the reading process we count the relative frequencies $\tilde{q}_i$ of observations for each interval l_i. In order to determine the best table entry, we use the KL divergence, (explained in chapter 3, section 3.4, equation (3.4)). In the final step of the estimation procedure, we determine the distribution of q_i from the look-up table that minimizes the KL divergence for the observed relative frequencies and select the corresponding LLR values.

Table 6.2: Example of a look-up table with probability distribution q_i and LLR values for MSB pages

R3											
q_0	q_1	q_2	q_3	q_4	q_5	l_0	l_1	l_2	l_3	l_4	l_5
0.478	0.018	0.005	0.005	0.031	0.464	-7.50	-5.75	2.00	2.00	6.25	7.50
0.476	0.018	0.005	0.005	0.018	0.477	-7.50	-5.25	-2.00	1.50	5.00	7.50
0.463	0.028	0.010	0.010	0.033	0.456	-7.50	-4.50	-1.50	1.75	5.00	7.50
0.471	0.022	0.009	0.012	0.035	0.451	-7.50	-3.75	-0.75	2.25	5.25	7.50
0.450	0.036	0.013	0.010	0.028	0.463	-7.50	-4.75	-2.00	1.00	4.50	7.50
R7											
0.487	0.012	0.003	0.004	0.013	0.481	-7.50	-3.50	-0.50	3.00	6.50	7.50
0.481	0.015	0.004	0.003	0.010	0.486	-7.50	-4.25	-1.50	1.75	5.25	7.50
0.479	0.017	0.004	0.002	0.007	0.490	-7.50	-5.00	-2.00	1.25	5.00	7.50
0.471	0.022	0.009	0.009	0.019	0.469	-7.50	-3.00	-1.00	1.75	4.50	7.50
0.471	0.023	0.010	0.011	0.023	0.463	-7.50	-3.25	-0.50	2.00	4.50	7.50
0.463	0.028	0.011	0.009	0.020	0.470	-7.50	-3.50	-1.25	1.25	4.00	7.50
0.451	0.035	0.013	0.009	0.016	0.476	-7.50	-4.00	-1.75	0.75	3.50	7.50

The main issue of this channel estimation procedure is finding a suitable set of representatives for each reference threshold. Note, that the LLR values are typically represented with a small number of bits, e.g. three or five bits per LLR value [76]. Due to the low resolution of the reliability information, we assume that only a small number of different representatives are required which are obtained by k-means clustering. Given a set of observations $z_1, \ldots, z_n$, where each observation z_i is a (2V+2)-dimensional real vector $z_i = (q_0, q_1, \ldots, q_V, l_0, l_1, \ldots, l_V)$, the k-means clustering aims to partition the observations into k sets $\mathcal{S} = \{\mathcal{S}_1, \mathcal{S}_2, \ldots, \mathcal{S}_k\}$. The objective of the k-mean algorithm is to find the partition

$$\underset{\mathcal{S}}{\operatorname{argmin}} \sum_{i=1}^{k} \sum_{z \in \mathcal{S}_i} |\omega(z - \hat{z}_i)^T|^2, \tag{6.20}$$

where $\hat{z}_i$ is the representative of S_i and ω is a vector of weights. The vector $\omega = (\omega_0, \ldots, \omega_{2V+1})$ is required, because the magnitudes of the elements of the vectors z_i vary significantly. In order to equalize the magnitudes, we use ω_i as the inverse of the mean of the ith element of the data vector, where the mean is taken over all observations, hence we have $\mathbb{E}(\omega z^T) = 2V + 2$, where $\mathbb{E}(\cdot)$ denotes the expectation operator.

In order to determine the size of the set of representatives, we run the k-means algorithm iteratively with different values of $k = \{2, 3, \ldots\}$. The search stops once the algorithm returns a set of representatives, where the two LLR values with the smallest magnitudes (e.g. ι_2 and ι_3 in Table 6.2) are not unique among all representatives. The result of the algorithm is the output of the k-means algorithm of the last round where the two smallest magnitudes are unique for all representatives.

For instance, the entries in Table 6.2 were found for a TLC from vendor-A. Table 6.2 presents R_3 and R_7 threshold of the MSB page, where a set of 80 different measurements per reference threshold was considered with 8 bits resolution for the LLR values. These measurements considered four different flash chips of the same flash type, different numbers of P/E cycles, and different endurance conditions. For this data set, the proposed algorithm returned 5 to 8 representatives per reference threshold, where the number of representatives increases for higher reference voltages. The values of the representatives vary significantly for the different reference thresholds.

6.4.2 Simulation results based on channel estimation

Now, we consider the performance of the channel estimation, i.e. the conditional probability of observing $0's$ or $1's$ in each interval. The simulation results are based on empirical data from a TLC designed for the EOL cases. Note that we consider extended BCH codes with length $n_a = 256$ for the simulation.

Example 9. *Consider the code from Example 6 from the previous chapter. This code has a code rate $R = 0.90$ and was designed according to error probability $\epsilon = 0.0066$. This code guarantees $P_e \leq 10^{-16}$ for $\epsilon \leq 0.0066$ according to Equation (5.35) in chapter 5. Note that this code is also able to correct burst errors. The minimum distance of all outer RS codes is greater than or equal* to three. Hence, each outer code can correct at least two erroneous symbols and consequently two columns of the codeword matrix may be corrupted by an arbitrary number of errors.

Table 6.3: Simulation results for a GCC with perfect channel information

level	page	perfect channel information for Vandor-A			
		BER	WER BCH	erasure BCH	WER RS (R=0.90)
0	LSB	0.0058	0.0279	0.221	1.62e-23
	CSB	0.0045	0.0134	0.249	2.47e-27
	MSB	**0.0061**	**0.0288**	**0.262**	**5.75e-20**
1	LSB	0.0058	9.5e-4	0.069	9.95e-29
	CSB	0.0045	2.2e-3	0.035	2.28e-31
	MSB	**0.0061**	**2.2e-3**	**0.084**	**7.65e-20**
2	LSB	0.0058	1.1e-4	0.009	1.33e-26
	CSB	0.0045	5.2e-5	0.006	1.47e-31
	MSB	**0.0061**	**2.1e-4**	**0.017**	**6.42e-19**

Table 6.2 contains two thresholds R_3 and R_7 of the MSB page. The simulated scenarios for the MSB correspond exactly to table entries of the LLR. Note that we do not present the table for other pages, i.e. the LSB and the CSB. Table 6.3 presents simulation results for the first three levels of the GCC, i.e. extended BCH codes of minimum Hamming distance $d_{b,0} = 4$, $d_{b,1} = 6$, and $d_{b,2} = 8$, respectively. The RS codes have minimum Hamming distance $d_{a,0} = 33$, $d_{a,1} = 89$, and $d_{a,1} = 115$, respectively. Table 6.3 illustrates the residual error probabilities with perfect channel knowledge. The third column presents the worst BER cases for different pages, i.e. the highest BER among 80 sets. For this simulation, we consider all pages, LSB, CSB and MSB. The next columns contain WER and erasure probabilities for the inner BCH codes, as well as the WER probabilities for the RS codes. For GCC codes, we consider code rates, $R = 0.90$. Note that the residual error probabilities of the GCC in worst cases achieve less than 10^{-16} with a perfect channel information.

Table. 6.4 presents the simulation results of the estimated channel information for MSB. Using the LLR values in Table 6.2, the estimation errors lead to a small degradation of the decoding performance as it can be observed by comparing Table 6.4 with the corresponding results from Table 6.3. Figure 6.6

Table 6.4: Simulation results for a GCC with estimated channel information

		estimated channel information Vandor-A			
level	page	BER	WER BCH	erasure BCH	WER RS (R=0.90)
0	MSB	0.0061	0.0255	0.2759	6.29e-20
1	MSB	·0.0061	0.0016	0.0907	2.42e-19

depicts histograms for the actually selected table entries. With this scenario the best table entry is selected in about 99% of all cases.

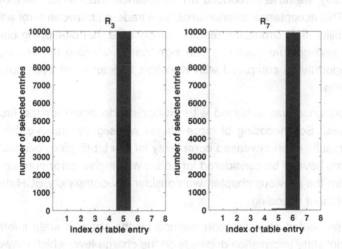

Figure 6.6: Histogram of selected table entries of MSB page (left for lower reference threshold R_3 with 5 entries, right for higher reference threshold R_7 with 7 entries).

6.5 Summary

Soft-input decoding algorithms can significantly improve the reliability of flash memories. However, an error correction scheme for flash memories also requires a good hard-input decoding performance. The threshold voltage sensing operation to obtain the soft information causes a larger energy consumption and a latency penalty. Hence, the soft-input decoding is only used for blocks where the hard-input decoding fails.

In this chapter, we have presented a soft-input decoder for high-rate GCC. The soft-decoding approach combines a bit-flipping strategy with algebraic decoding. Hence, the decoder components for the hard-input decoding can be utilized and the additional overhead for the soft-input decoding is relatively low, because only a small number of test patterns is required.

In order to exploit the error and erasure decoding of the outer RS codes for soft-decoding, we have introduced an acceptance criterion for the bit-flipping decoder. This acceptance criterion enables a trade-off between error and erasure probability. The simulation results demonstrate that bit-flipping combined with error and erasure decoding can significantly improve the soft-input decoding performance compared with Kaneko's rule and can even outperform ML decoding.

This performance was achieved with bit-flipping decoding for the first three code stages. Soft-decoding of more stages will slightly improve the coding gain, but results in an increased complexity for the bit-flipping, because more test patterns have to be considered for codes with higher error-correcting capability. From the previous chapter, we consider low-complexity BCH decoding for the soft-input decoding.

We have presented an estimation method for the channel state information. The channel state information depends on the charge level which may change due to P/E cycles, data retention, read disturb and other effects. The proposed channel model considers different voltage distributions for different charge levels. The different distributions model the asymmetric error characteristic of MLC and TLC flash memories. The proposed estimation method for the channel state information is solely based on counting the number of observations in the different quantization intervals. Hence, it does not require additional pilot data for channel estimation. Moreover, page-wise read operations can be supported. The presented simulation results for GCC indicate that the LLR values are estimated with sufficient accuracy.

7 Conclusions

This dissertation has proposed various applicable techniques to improve the reliability of NAND flash memory. In summary, this dissertation makes the following contributions:

- We propose an offset calibration method that aims to minimize the BER by adapting the read threshold voltages. This low-latency read-retry mechanism is based on meta-data, which is protected by strong ECC. Based on a few hundred bits of meta-data, the proposal mechanism achieves BER values that are close to the results with the optimal thresholds. This can be obtained only with two reading operations for the worst cases, i.e. at the EOL.

- We propose adaptive universal data compression algorithms that are applicable for flash memories. The proposed scheme is a combination of the MTF algorithm followed by Huffman coding. This scheme results in low complexity and latency compared with LZW and PDLZW. Considering the WA effect, the proposed scheme achieves the same performance compared with the PDLZW.

  In order to improve the compression performance, we integrate the BWT algorithm to MH scheme. This results in a significant compression performance but requires more latency compared with MH scheme. This latency can be reduced by minimizing the block sorting size of the BWT algorithm. On average, the WA reduces to less than half for a particular data model. This improves flash endurance, where this scheme outperforms the LZW and PDLZW schemes.

- MLC and TLC flash memories require a channel model that reflects the asymmetric error characteristic of the flash channel. In this respect, we propose a combined coding scheme that combines data compression and ECC. The data compression reduces the data size to increase the error correction capability, i.e improving the reliability of MLC and TLC flash memories. For the investigated TLC flash, the proposed coding scheme can improve the life time by up to 6000 P/E cycles compared with uncompressed data, and 5000 P/E cycles compared with LZW and PDLZW schemes.

- We construct the GCC for high code rate for flash memories. The GCCs are constructed from inner BCH codes and outer RS codes. Applying SPC code in the first level of the inner code enables high-rate codes, and low complexity compared with long BCH codes. Furthermore, the single- and double-extended BCH codes provide a significant performance that outperforms the HPC in [109]. The GCC based on double-extended code achieves WER less than 10^{-16} at SNR $= 5.3$ dB. Moreover, the hard-input decoding achieves the same performance of the soft-input decoding in [144], which is based on sequential stack decoding of GCC.

- Finally, we propose a soft-input decoder for GCC. The inner BCH codes are decoded with a bit-flipping decoding. We adapt bit-flipping decoding for the coarsely-quantized soft information of the flash channel. An acceptance criterion which determines the final codeword outperforms Kaneko's rule and ML decoding in combination with outer error and erasure decoding. The proposed scheme outperforms the hard-input BCH decoding, hard-input GCC decoding, and soft-input GCC decoding in [144] by about 1.3 dB, 0.5 dB, and 0.4 dB, respectively.

Moreover, we have proposed a channel estimation approach for flash memories that avoids additional pilot data. Based on flash measurements, soft-input decoding with estimated channel information achieves almost the same performance compared with the perfect channel information. Low-complexity soft-input decoding are guaranteed to achieve WER less than 10^{-16} for TLC at the EOL.

Bibliography

[1] J. Freudenberger, M. Rajab, and S. Shavgulidze, "A source and channel coding approach for improving flash memory endurance," *IEEE Transactions on Very Large Scale Integration (VLSI) Systems*, vol. 26, no. 5, pp. 981–990, May 2018.

[2] J. Freudenberger, M. Rajab, and S. Shavgulidze, "A channel and source coding approach for the binary asymmetric channel with applications to MLC flash memories," in *11th International ITG Conference on Systems, Communications and Coding (SCC), Hamburg*, Feb. 2017, pp. 1–4.

[3] J. Freudenberger, M. Rajab, and S. Shavgulidze, "Estimation of channel state information for non-volatile flash memories," in *IEEE 7th International Conference on Consumer Electronics (ICCE)*, Sept 2017.

[4] M. Rajab, J. Thiers, and J. Freudenberger, "Read threshold calibration for non-volatile flash memories," in *IEEE 9th International Conference on Consumer Electronics (ICCE)*, Sept 2019.

[5] Alessandro S. Spinelli, Christian Monzio Compagnoni, and Andrea L. Lacaita, "Reliability of NAND flash memories: Planar cells and emerging issues in 3D devices," *Computers*, vol. 6, no. 2, 2017.

[6] Ralph Howard Fowler and Lothar Nordheim, "Electron emission in intense electric fields," *Proc. R. Soc. Lond. A*, vol. 119, no. 781, pp. 173–181, 1928.

[7] Yu Cai, Yixin Luo, Saugata Ghose, and Onur Mutlu, "Read disturb errors in MLC NAND flash memory: Characterization, mitigation, and recovery," in *45th Annual IEEE/IFIP International Conference on Dependable Systems and Networks (DSN)*. IEEE, 2015, pp. 438–449.

[8] V. Taranalli, H. Uchikawa, and P. H. Siegel, "Channel models for multilevel cell flash memories based on empirical error analysis," *IEEE Transactions on Communications*, vol. 64, no. 8, pp. 3169 – 3181, Aug 2016.

© The Editor(s) (if applicable) and The Author(s), under exclusive license to
Springer Fachmedien Wiesbaden GmbH, part of Springer Nature 2020
M. Rajab, *Channel and Source Coding for Non-Volatile Flash Memories*,
Schriftenreihe der Institute für Systemdynamik (ISD) und optische
Systeme (IOS), https://doi.org/10.1007/978-3-658-28982-9

[9] Seiji Yamada, Youhei Hiura, Tomoko Yamane, Kazumi Amemiya, Yoichi Ohshima, and K Yoshikawa, "Degradation mechanism of flash eeprom programming after program/erase cycles," in *Electron Devices Meeting, 1993. IEDM'93. Technical Digest., International.* IEEE, 1993, pp. 23–26.

[10] P. Mora, S. Renard, G. Bossu, P. Waltz, G. Pananakakis, and G. Ghibaudo, "Reliability issues related to fast charge loss mechanism in embedded non volatile memories," in *IEEE International Integrated Reliability Workshop*, Oct 2006, pp. 68–72.

[11] Yu Cai, Yixin Luo, Erich F Haratsch, Ken Mai, and Onur Mutlu, "Data retention in mlc nand flash memory: Characterization, optimization, and recovery," in *High Performance Computer Architecture (HPCA), 2015 IEEE 21st International Symposium on.* IEEE, 2015, pp. 551–563.

[12] D. Oh, B. Lee, E. Kwon, S. Kim, G. Cho, S. Park, S. Lee, and S. Hong, "TCAD simulation of data retention characteristics of charge trap device for 3-D NAND flash memory," in *IEEE International Memory Workshop (IMW)*, May 2015, pp. 1–4.

[13] Frederic Sala, Ryan Gabrys, and Lara Dolecek, "Dynamic threshold schemes for multi-level non-volatile memories," *IEEE Transactions on Communications*, vol. 61, no. 7, pp. 2624–2634, 2013.

[14] Hongchao Zhou, Anxiao Jiang, and Jehoshua Bruck, "Error-correcting schemes with dynamic thresholds in nonvolatile memories," in *Information Theory Proceedings (ISIT), 2011 IEEE International Symposium on.* IEEE, 2011, pp. 2143–2147.

[15] C. H. Chan, Y. Zhu, U. F. Chio, S. W. Sin, U. Seng-Pan, and R. P. Martins, "A threshold-embedded offset calibration technique for inverter-based flash adcs," in *2010 53rd IEEE International Midwest Symposium on Circuits and Systems*, Aug 2010, pp. 489–492.

[16] Nikolaos Papandreou, Thomas Parnell, Haralampos Pozidis, Thomas Mittelholzer, Evangelos Eleftheriou, Charles Camp, Thomas Griffin, Gary Tressler, and Andrew Walls, "Using adaptive read voltage thresholds to enhance the reliability of mlc nand flash memory systems," in *Proceedings of the 24th edition of the great lakes symposium on VLSI.* ACM, 2014, pp. 151–156.

[17] Jae-Duk Lee, Sung-Hoi Hur, and Jung-Dal Choi, "Effects of floating-gate interference on nand flash memory cell operation," *IEEE Electron Device Letters*, vol. 23, no. 5, pp. 264–266, 2002.

[18] Yu Cai, Gulay Yalcin, Onur Mutlu, Erich F Haratsch, Osman Unsal, Adrian Cristal, and Ken Mai, "Neighbor-cell assisted error correction for mlc nand flash memories," in *ACM SIGMETRICS Performance Evaluation Review*. ACM, 2014, vol. 42, pp. 491–504.

[19] Ki-Tae Park, Myounggon Kang, Doogon Kim, Soon-Wook Hwang, Byung Yong Choi, Yeong-Taek Lee, Changhyun Kim, and Kinam Kim, "A zeroing cell-to-cell interference page architecture with temporary lsb storing and parallel msb program scheme for mlc nand flash memories," *IEEE Journal of Solid-State Circuits*, vol. 43, no. 4, pp. 919–928, 2008.

[20] C. J. Wu, H. T. Lue, T. H. Hsu, C. C. Hsieh, W. C. Chen, P. Y. Du, C. J. Chiu, and C. Y. Lu, "Device characteristics of single-gate vertical channel (SGVC) 3D NAND flash architecture," in *IEEE 8th International Memory Workshop (IMW)*, May 2016, pp. 1–4.

[21] H. Li, "Modeling of threshold voltage distribution in NAND flash memory: A monte carlo method," *IEEE Transactions on Electron Devices*, vol. 63, no. 9, pp. 3527–3532, Sept 2016.

[22] Paolo Cappelletti, Roberto Bez, Daniele Cantarelli, and Lorenzo Fratin, "Failure mechanisms of flash cell in program/erase cycling," in *Electron Devices Meeting, 1994. IEDM'94. Technical Digest., International*. IEEE, 1994, pp. 291–294.

[23] H. Wang, T. Y. Chen, and R. D. Wesel, "Histogram-based flash channel estimation," in *IEEE International Conference on Communications (ICC)*, June 2015, pp. 283–288.

[24] Tsung-Yi Chen, Adam R Williamson, and Richard D Wesel, "Increasing flash memory lifetime by dynamic voltage allocation for constant mutual information," in *Information Theory and Applications Workshop (ITA), 2014*. IEEE, 2014, pp. 1–5.

[25] Nikolaos Papandreou, Thomas Parnell, Haralampos Pozidis, Thomas Mittelholzer, Evangelos Eleftheriou, Charles Camp, Thomas Griffin, Gary Tressler, and Andrew Walls, "Using adaptive read voltage

thresholds to enhance the reliability of MLC NAND flash memory systems," in *Proceedings of the 24th Edition of the Great Lakes Symposium on VLSI*, New York, NY, USA, 2014, GLSVLSI '14, pp. 151–156, ACM.

[26] Borja Peleato, Rajiv Agarwal, John M Cioffi, Minghai Qin, and Paul H Siegel, "Adaptive read thresholds for NAND flash," *IEEE Transactions on Communications*, vol. 63, no. 9, pp. 3069–3081, 2015.

[27] Y. Cai, E. F. Haratsch, O. Mutlu, and K. Mai, "Threshold voltage distribution in MLC NAND flash memory: Characterization, analysis, and modeling," in *2013 Design, Automation Test in Europe Conference Exhibition (DATE)*, March 2013, pp. 1285–1290.

[28] Youngjo Park and Jin-Soo Kim, "zFTL: power-efficient data compression support for NAND flash-based consumer electronics devices," *IEEE Transactions on Consumer Electronics*, vol. 57, no. 3, pp. 1148–1156, August 2011.

[29] Ningde Xie, Guiqiang Dong, and Tong Zhang, "Using lossless data compression in data storage systems: Not for saving space," *IEEE Transactions on Computers*, vol. 60, no. 3, pp. 335–345, March 2011.

[30] M. Martina, C. Condo, G. Masera, and M. Zamboni, "A joint source/channel approach to strengthen embedded programmable devices against flash memory errors," *IEEE Embedded Systems Letters*, vol. 6, no. 4, pp. 77–80, Dec 2014.

[31] Jiangpeng Li, Kai Zhao, Xuebin Zhang, Jun Ma, Ming Zhao, and Tong Zhang, "How much can data compressibility help to improve NAND flash memory lifetime?," in *Proceedings of the 13th USENIX Conference on File and Storage Technologies (FAST15)*, Feb. 2015, pp. 227–240.

[32] J. Freudenberger, A. Beck, and M. Rajab, "A data compression scheme for reliable data storage in non-volatile memories," in *IEEE 5th International Conference on Consumer Electronics (ICCE)*, Sept 2015, pp. 139–142.

[33] T. Ahrens, M. Rajab, and J. Freudenberger, "Compression of short data blocks to improve the reliability of non-volatile flash memories," in *International Conference on Information and Digital Technologies (IDT)*, July 2016, pp. 1–4.

[34] Jon Louis Bentley, Daniel D. Sleator, Robert E. Tarjan, and Victor K. Wei, "A locally adaptive data compression scheme," *Commun. ACM*, vol. 29, no. 4, pp. 320–330, Apr. 1986.

[35] P. Elias, "Interval and recency rank source coding: Two on-line adaptive variable-length schemes," *IEEE Transactions on Information Theory*, vol. 33, no. 1, pp. 3–10, Jan 1987.

[36] F.M.J. Willems, "Universal data compression and repetition times," *IEEE Transactions on Information Theory*, vol. 35, no. 1, pp. 54–58, Jan 1989.

[37] D. A. Huffman, "A method for the construction of minimum-redundancy codes," *Proceedings of the IRE*, vol. 40, no. 9, pp. 1098–1101, Sept 1952.

[38] Jossy Sayir, Ivo Spieler, and Patricia Portmann, "Conditional recency-ranking for source coding," in *Proc. IEEE Information Theory Workshop*, June 1996, p. 61.

[39] B. Balkenhol, S. Kurtz, and Y.M. Shtarkov, "Modifications of the Burrows and Wheeler data compression algorithm," in *Proceedings Data Compression Conference (DCC99)*, Mar 1999, pp. 188–197.

[40] M. Burrows and D. Wheeler, *A block-sorting lossless data compression algorithm*, SRC Research Report 124, Digital Systems Research Center, Palo Alto, CA., 1994.

[41] T.A. Welch, "A technique for high-performance data compression," *Computer*, vol. 17, no. 6, pp. 8–19, June 1984.

[42] Ming-Bo Lin, Jang-Feng Lee, and Gene Eu Jan, "A lossless data compression and decompression algorithm and its hardware architecture," *IEEE Transactions on Very Large Scale Integration (VLSI) Systems*, vol. 14, no. 9, pp. 925–936, Sept 2006.

[43] J. Freudenberger, M. Rajab, D. Rohweder, and M. Safieh, "A codec architecture for the compression of short data blocks," *Journal of Circuits, Systems, and Computers (JCSC)*, vol. 27, no. 2, pp. 1–17, Feb. 2018.

[44] Peter Elias, "Universal codeword sets and representations of the integers," *IEEE transactions on information theory*, vol. 21, no. 2, pp. 194–203, 1975.

[45] James L Massey, "Applied digital information theory," *lecture notes, ETH Zurich.[Online]. Available: http://www. isiweb. ee. ethz. ch/archive/massey scr/adit1. pdf*, 1980.

[46] P. M. Szecowka and T. Mandrysz, "Towards hardware implementation of bzip2 data compression algorithm," in *16th International Conference Mixed Design of Integrated Circuits Systems (MIXDES)*, June 2009, pp. 337–340.

[47] T.C. Bell, J.G. Cleary, and I.H. Witten, *Text compression*, Prentice Hall, Englewood Cliffs, NJ, 1990.

[48] Matt Powell, "Evaluating lossless compression methods," in *New Zealand Computer Science Research Students' Conference, Canterbury*, 2001, pp. 35–41.

[49] M. Gutman, "Fixed-prefix encoding of the integers can be Huffman-optimal," *IEEE Transactions on Information Theory*, vol. 36, no. 4, pp. 936–938, Jul 1990.

[50] S. Kullback, *Information theory and statistics*, John Wiley & Sons, 1959.

[51] Claude Elwood Shannon, "A mathematical theory of communication," *Bell system technical journal*, vol. 27, no. 3, pp. 379–423, 1948.

[52] B. Ash Robert, *Information Theory*, Dover Publications Inc., New York, 1990.

[53] Jacob Ziv and Abraham Lempel, "A universal algorithm for sequential data compression," *IEEE Transactions on information theory*, vol. 23, no. 3, pp. 337–343, 1977.

[54] R. Samanta and R. N. Mahapatra, "An enhanced CAM architecture to accelerate LZW compression algorithm," in *20th International Conference on VLSI Design held jointly with 6th International Conference on Embedded Systems (VLSID'07)*, Jan 2007, pp. 824–829.

[55] J Jiang and S Jones, "Word-based dynamic algorithms for data compression," *IEE Proceedings I (Communications, Speech and Vision)*, vol. 139, no. 6, pp. 582–586, 1992.

[56] Ming-Bo Lin, "A hardware architecture for the LZW compression and decompression algorithms based on parallel dictionaries," *Journal of*

VLSI signal processing systems for signal, image and video technology, vol. 26, no. 3, pp. 369–381, Nov. 2000.

[57] M. Schindler, "A fast block-sorting algorithm for lossless data compression," in *Data Compression Conference*, March 1997, pp. 469–.

[58] Malek Safieh and Jürgen Freudenberger, "Pipelined decoder for the limited context order burrows-wheeler-transformation," *IET Circuits, Devices & Systems*, 2018.

[59] J. Martinez, R. Cumplido, and C. Feregrino, "An FPGA-based parallel sorting architecture for the Burrows Wheeler transform," in *International Conference on Reconfigurable Computing and FPGAs*, Sept 2005, pp. 1–7.

[60] S. Arming, R. Fenkhuber, and T. Handl, "Data compression in hardware – the Burrows-Wheeler approach," in *13th IEEE Symposium on Design and Diagnostics of Electronic Circuits and Systems*, April 2010, pp. 60–65.

[61] U. I. Cheema and A. A. Khokhar, "A high performance architecture for computing Burrows-Wheeler transform on FPGAs," in *International Conference on Reconfigurable Computing and FPGAs*, Dec 2013, pp. 1–6.

[62] S. Hayashi and K. Taura, "Parallel and memory-efficient Burrows-Wheeler transform," in *2013 IEEE International Conference on Big Data*, Oct 2013, pp. 43–50.

[63] S. Dong, X. Wang, and X. Wang, "A novel high-speed parallel scheme for data sorting algorithm based on FPGA," in *2009 2nd International Congress on Image and Signal Processing*, Oct 2009, pp. 1–4.

[64] Malek Safieh and Jürgen Freudenberger, "Efficient vlsi architecture for the parallel dictionary lzw data compression algorithm," *IET Circuits, Devices & Systems*, 2019.

[65] R. Micheloni, A. Marelli, and R. Ravasio, *Error Correction Codes for Non-Volatile Memories*, Springer, 2008.

[66] Congming Gao, Liang Shi, Kaijie Wu, C.J. Xue, and E.H.-M. Sha, "Exploit asymmetric error rates of cell states to improve the performance of

flash memory storage systems," in *Computer Design (ICCD), 2014 32nd IEEE International Conference on*, Oct 2014, pp. 202–207.

[67] Martin Bossert, *Channel coding for telecommunications*, Wiley, 1999.

[68] A. Neubauer, J. Freudenberger, and V. Kühn, *Coding Theory: Algorithms, Architectures and Applications*, John Wiley & Sons, 2007.

[69] F. Sun, S. Devarajan, K. Rose, and T. Zhang, "Design of on-chip error correction systems for multilevel NOR and NAND flash memories," *IET Circuits, Devices Systems*, vol. 1, no. 3, pp. 241 –249, June 2007.

[70] Guiqiang Dong, Ningde Xie, and Tong Zhang, "On the use of soft-decision error-correction codes in NAND Flash memory," *IEEE Transactions on Circuits and Systems I: Regular Papers*, vol. 58, no. 2, pp. 429–439, Feb 2011.

[71] Wei Liu, Junrye Rho, and Wonyong Sung, "Low-power high-throughput BCH error correction VLSI design for multi-level cell NAND flash memories," in *IEEE Workshop on Signal Processing Systems Design and Implementation (SIPS)*, oct. 2006, pp. 303 –308.

[72] J. Freudenberger and J. Spinner, "A configurable Bose-Chaudhuri-Hocquenghem codec architecture for flash controller applications," *Journal of Circuits, Systems, and Computers*, vol. 23, no. 2, pp. 1–15, Feb 2014.

[73] C. Yang, Y. Emre, and C. Chakrabarti, "Product code schemes for error correction in MLC NAND flash memories," *IEEE Transactions on Very Large Scale Integration (VLSI) Systems*, vol. 20, no. 12, pp. 2302–2314, Dec 2012.

[74] S. Li and T. Zhang, "Improving multi-level NAND flash memory storage reliability using concatenated BCH-TCM coding," *IEEE Transactions on Very Large Scale Integration (VLSI) Systems*, vol. 18, no. 10, pp. 1412–1420, Oct 2010.

[75] J. Oh, J. Ha, J. Moon, and G. Ungerboeck, "Rs-enhanced TCM for multilevel flash memories," *IEEE Transactions on Communications*, vol. 61, no. 5, pp. 1674–1683, May 2013.

[76] J. Spinner, J. Freudenberger, and S. Shavgulidze, "A soft input decoding algorithm for generalized concatenated codes," *IEEE Transactions on Communications*, vol. 64, no. 9, pp. 3585–3595, Sept 2016.

[77] I. Zhilin and A. Kreschuk, "Generalized concatenated code constructions with low overhead for optical channels and nand-flash memory," in *XV International Symposium Problems of Redundancy in Information and Control Systems (REDUNDANCY)*, Sept 2016, pp. 177–180.

[78] J. Spinner, M. Rajab, and J. Freudenberger, "Construction of high-rate generalized concatenated codes for applications in non-volatile flash memories," in *IEEE 8th International Memory Workshop (IMW)*, May 2016, pp. 1–4.

[79] Kai Zhao, Wenzhe Zhao, Hongbin Sun, Xiaodong Zhang, Nanning Zheng, and Tong Zhang, "LDPC-in-SSD: Making advanced error correction codes work effectively in solid state drives," in *Presented as part of the 11th USENIX Conference on File and Storage Technologies (FAST 13)*, San Jose, CA, 2013, pp. 243–256, USENIX.

[80] Jiadong Wang, K. Vakilinia, Tsung-Yi Chen, T. Courtade, Guiqiang Dong, Tong Zhang, H. Shankar, and R. Wesel, "Enhanced precision through multiple reads for LDPC decoding in flash memories," *IEEE Journal on Selected Areas in Communications*, vol. 32, no. 5, pp. 880–891, May 2014.

[81] K. Haymaker and C. A. Kelley, "Structured bit-interleaved LDPC codes for MLC flash memory," *IEEE Journal on Selected Areas in Communications*, vol. 32, no. 5, pp. 870–879, May 2014.

[82] E. Yaakobi, Jing Ma, L. Grupp, P.H. Siegel, S. Swanson, and J.K Wolf, "Error characterization and coding schemes for flash memories," in *IEEE GLOBECOM Workshops*, Dec. 2010, pp. 1856–1860.

[83] E. Yaakobi, L. Grupp, P.H. Siegel, S. Swanson, and J.K. Wolf, "Characterization and error-correcting codes for TLC flash memories," in *International Conference on Computing, Networking and Communications (ICNC)*, Jan 2012, pp. 486–491.

[84] R. Gabrys, E. Yaakobi, and L. Dolecek, "Graded bit-error-correcting codes with applications to flash memory," *IEEE Transactions on Information Theory*, vol. 59, no. 4, pp. 2315–2327, April 2013.

[85] R. Gabrys, F. Sala, and L. Dolecek, "Coding for unreliable flash memory cells," *IEEE Communications Letters*, vol. 18, no. 9, pp. 1491–1494, Sept 2014.

[86] L. Dolecek and Y. Cassuto, "Channel coding for nonvolatile memory technologies: Theoretical advances and practical considerations," *Proceedings of the IEEE*, vol. 105, no. 9, pp. 1705–1724, Sept 2017.

[87] Y. Cai, E. F. Haratsch, O. Mutlu, and K. Mai, "Error patterns in MLC NAND flash memory: Measurement, characterization, and analysis," in *Design, Automation Test in Europe Conference Exhibition (DATE)*, March 2012, pp. 521–526.

[88] M. Grassl, P. W. Shor, G. Smith, J. Smolin, and B. Zeng, "New constructions of codes for asymmetric channels via concatenation," *IEEE Transactions on Information Theory*, vol. 61, no. 4, pp. 1879–1886, April 2015.

[89] Shu Lin and Daniel J Costello, *Error control coding*, Pearson Education India, 2001.

[90] Florence Jessie MacWilliams and Neil James Alexander Sloane, *The theory of error-correcting codes*, North Holland Publishing Co., June 1988.

[91] X. Li, Z. Huo, L. Jin, Y. Wang, J. Liu, D. Jiang, X. Yang, and M. Liu, "Investigation of charge loss mechanisms in 3D TANOS cylindrical junctionless charge trapping memory," in *12th IEEE International Conference on Solid-State and Integrated Circuit Technology (ICSICT)*, Oct 2014, pp. 1–3.

[92] Y. Cai, Y. Luo, E. F. Haratsch, K. Mai, and O. Mutlu, "Data retention in MLC NAND flash memory: Characterization, optimization, and recovery," in *IEEE 21st International Symposium on High Performance Computer Architecture (HPCA)*, Feb 2015, pp. 551–563.

[93] G. Nong and S. Zhang, "Efficient algorithms for the inverse sort transform," *IEEE Transactions on Computers*, vol. 56, no. 11, pp. 1564–1574, 2007.

[94] Xinmiao Zhang and K K Parhi, "High-speed architectures for parallel long BCH encoders," *IEEE Transactions on Very Large Scale Integration (VLSI) Systems*, vol. 13, no. 7, pp. 872–877, 2005.

[95] J. Freudenberger, U. Kaiser, and J. Spinner, "Concatenated code constructions for error correction in non-volatile memories," in *Int. Symposium on Signals, Systems, and Electronics (ISSSE), Potsdam*, Oct 2012, pp. 1–6.

[96] P. Trifonov and Y. Wang, "Generalized concatenated codes for block and device failure protection," in *IEEE International Conference on Computer and Information Technology (CIT)*, Dec 2016, pp. 1–9.

[97] V. V. Zyablov I. V. Zhilin, "Generalized error-locating codes with component codes over the same alphabet," *Problems Inform. Transmission*, vol. 53, no. 2, pp. 114–135, Sept 2017.

[98] Chengen Yang, Yunus Emre, and Chaitali Chakrabarti, "Product code schemes for error correction in mlc nand flash memories," *IEEE Transactions on Very Large Scale Integration (VLSI) Systems*, vol. 20, no. 12, pp. 2302–2314, 2012.

[99] Changgeun Kim, Sunwook Rhee, Juhee Kim, and Yong Jee, "Product reed-solomon codes for implementing nand flash controller on fpga chip," in *Computer Engineering and Applications (ICCEA), 2010 Second International Conference on*. IEEE, 2010, vol. 1, pp. 281–285.

[100] Peter Elias, "Error-free coding," *Transactions of the IRE Professional Group on Information Theory*, vol. 4, no. 4, pp. 29–37, 1954.

[101] Christian Häger and Henry D Pfister, "Approaching miscorrection-free performance of product codes with anchor decoding," *IEEE Transactions on Communications*, 2018.

[102] Norman Abramson, "Cascade decoding of cyclic product codes," *IEEE Transactions on Communication Technology*, vol. 16, no. 3, pp. 398–402, 1968.

[103] Jorn Justesen, "Performance of product codes and related structures with iterated decoding," *IEEE Transactions on Communications*, vol. 59, no. 2, pp. 407–415, 2011.

[104] J. Freudenberger, M. Rajab, and S. Shavgulidze, "A low-complexity three-error-correcting bch decoder with applications in concatenated codes," in *SCC 2019; 12th International ITG Conference on Systems, Communications and Coding; Proceedings of.* VDE, 2019, pp. 1–6.

[105] Eirik Rosnes, "Stopping set analysis of iterative row-column decoding of product codes," *IEEE Transactions on Information Theory*, vol. 54, no. 4, pp. 1551–1560, 2008.

[106] Santosh Emmadi, Krishna R Narayanan, and Henry D Pfister, "Half-product codes for flash memory," in *Proc. Non-Volatile Memories Workshop*, 2015, pp. 1–2.

[107] Henry D Pfister, Santosh K Emmadi, and Krishna Narayanan, "Symmetric product codes," in *Information Theory and Applications Workshop (ITA), 2015.* IEEE, 2015, pp. 282–290.

[108] Sung-gun Cho, Daesung Kim, Jinho Choi, and Jeongseok Ha, "Block-wise concatenated bch codes for nand flash memories," *IEEE Transactions on Communications*, vol. 62, no. 4, pp. 1164–1177, 2014.

[109] Daesung Kim, Krishna Narayanan, and Jeongseok Ha, "Symmetric block-wise concatenated bch codes for nand flash memories," *IEEE Transactions on Communications*, 2018.

[110] S. Cho, Daesung Kim, Jinho Choi, and Jeongseok Ha, "Block-wise concatenated BCH codes for NAND flash memories," *IEEE Transactions on Communications*, vol. 62, no. 4, pp. 1164–1177, April 2014.

[111] Daesung Kim and Jeongseok Ha, "Quasi-primitive block-wise concatenated BCH codes for NAND flash memories," in *IEEE Information Theory Workshop (ITW)*, Nov 2014, pp. 611–615.

[112] D. Kim and J. Ha, "Quasi-primitive block-wise concatenated BCH codes with collaborative decoding for NAND flash memories," *IEEE Transactions on Communications*, vol. 63, no. 10, pp. 3482–3496, Oct 2015.

[113] D. Kim and J. Ha, "Serial quasi-primitive BC-BCH codes for NAND flash memories," in *2016 IEEE International Conference on Communications (ICC)*, May 2016, pp. 1–6.

[114] K. Lee, H. Kang, J. Park, and H. Lee, "100Gb/s two-iteration concatenated BCH decoder architecture for optical communications," in *2010 IEEE Workshop On Signal Processing Systems*, Oct 2010, pp. 404–409.

[115] X. Zhang and Z. Wang, "A low-complexity three-error-correcting BCH decoder for optical transport network," *IEEE Transactions on Circuits and Systems II: Express Briefs*, vol. 59, no. 10, pp. 663–667, Oct 2012.

[116] J. Spinner and J. Freudenberger, "Decoder architecture for generalized concatenated codes," *IET Circuits, Devices & Systems*, vol. 9, no. 5, pp. 328–335, 2015.

[117] J. Spinner, D. Rohweder, and J. Freudenberger, "Soft input decoder for high-rate generalised concatenated codes," *IET Circuits, Devices Systems*, vol. 12, no. 4, pp. 432–438, 2018.

[118] B. P. Smith, A. Farhood, A. Hunt, F. R. Kschischang, and J. Lodge, "Staircase codes: FEC for 100 Gb/s OTN," *Journal of Lightwave Technology*, vol. 30, no. 1, pp. 110–117, Jan 2012.

[119] G. Hu, J. Sha, and Z. Wang, "Beyond 100Gbps encoder design for staircase codes," in *2016 IEEE International Workshop on Signal Processing Systems (SiPS)*, Oct 2016, pp. 154–158.

[120] D. Strukov, "The area and latency tradeoffs of binary bit-parallel BCH decoders for prospective nanoelectronic memories," in *2006 Fortieth Asilomar Conference on Signals, Systems and Computers*, Oct 2006, pp. 1183–1187.

[121] P. Amato, C. Laurent, M. Sforzin, S. Bellini, M. Ferrari, and A. Tomasoni, "Ultra fast, two-bit ECC for emerging memories," in *2014 IEEE 6th International Memory Workshop (IMW)*, May 2014, pp. 1–4.

[122] C. Yang, M. Mao, Y. Cao, and C. Chakrabarti, "Cost-effective design solutions for enhancing PRAM reliability and performance," *IEEE Transactions on Multi-Scale Computing Systems*, vol. 3, no. 1, pp. 1–11, Jan 2017.

[123] Erl-Huei Lu, Shao-Wei Wu, and Yi-Chang Cheng, "A decoding algorithm for triple-error-correcting binary BCH codes," *Information Processing Letters*, vol. 80, no. 6, pp. 299 – 303, 2001.

[124] G David Forney, "Concatenated codes," 1965.

[125] Victor Zyablov, Sergo Shavgulidze, and Martin Bossert, "An introduction to generalized concatenated codes," *European Transactions on Telecommunications*, vol. 10, no. 6, pp. 609–622, 1999.

[126] JEDEC Standard JESD218, "Solid-state drive (ssd) requirements and endurance test method," *Arlington, VA, JEDEC Solid State Technology Association*, vol. 1, 2010.

[127] W. Peterson, "Encoding and error-correction procedures for the Bose-Chaudhuri codes," *IRE Transactions on Information Theory*, vol. 6, no. 4, pp. 459–470, September 1960.

[128] Shu Lin and Daniel J. Costello, *Error Control Coding*, Upper Saddle River, NJ: Prentice-Hall, 2004.

[129] Yuan Jiang, *A practical guide to error-control coding using Matlab*, Artech House, 2010.

[130] N. Ahmadi, M. H. Sirojuddiin, A. D. Nandaviri, and T. Adiono, "An optimal architecture of BCH decoder," in *2010 4th International Conference on Application of Information and Communication Technologies*, Oct 2010, pp. 1–5.

[131] Toshiya Itoh and Shigeo Tsujii, "A fast algorithm for computing multiplicative inverses in GF(2^m) using normal bases," *Information and computation*, vol. 78, no. 3, pp. 171–177, 1988.

[132] I. Dumer, *Concatenated codes and their multilevel generalizations*, in Handbook of Coding Theory, Vol. II, Elsevier, Amsterdam, 1998.

[133] L. Weiburn and J.K. Cavers, "Improved performance of Reed-Solomon decoding with the use of pilot signals for erasure generation," in *Vehicular Technology Conference, 1998. VTC 98. 48th IEEE*, May 1998, vol. 3, pp. 1930–1934 vol.3.

[134] U. Wachsmann, R.F.H. Fischer, and J.B. Huber, "Multilevel codes: theoretical concepts and practical design rules," *IEEE Transactions on Information Theory*, vol. 45, no. 5, pp. 1361–1391, Jul 1999.

[135] I. Zhilin, A. Kreschuk, and V. Zyablov, "Generalized concatenated codes with soft decoding of inner and outer codes," in *International Symposium on Information Theory and Its Applications (ISITA)*, Oct 2016, pp. 290–294.

[136] J. Freudenberger, M. Rajab, and S. Shavgulidze, "A soft-input bit-flipping decoder for generalized concatenated codes," in *2018 IEEE International Symposium on Information Theory (ISIT)*, June 2018, pp. 1301–1305.

[137] R. Micheloni, A. Marelli, and R. Ravasio, *Error Correction Codes for Non-Volatile Memories*, Springer, 2008.

[138] M. Rajab, J. Freudenberger, and S. Shavgulidze, "Soft-input bit-flipping decoding of generalized concatenated codes for application in nonvolatile flash memories," *IET Communications*, Nov, 2018.

[139] J. Freudenberger and M. Rajab, "Chase decoding for quantized reliability information with applications to flash memories," *BW-CAR| SINCOM*, p. 7, 2016.

[140] J. Freudenberger, M. Rajab, and C. Baumhof, "Methods and apparatus for error correction coding based on data compression," June 2018, US Patent App. 15/848,012.

[141] Wei Lin, Shao-Wei Yen, Yu-Cheng Hsu, Yu-Hsiang Lin, Li-Chun Liang, Tien-Ching Wang, Pei-Yu Shih, Kuo-Hsin Lai, Kuo-Yi Cheng, and Chun-Yen Chang, "A low power and ultra high reliability LDPC error correction engine with digital signal processing for embedded NAND flash controller in 40nm CMOS," in *Symposium on VLSI Circuits Digest of Technical Papers*, June 2014, pp. 1–2.

[142] J. Guo, W. Wen, J. Hu, D. Wang, H. Li, and Y. Chen, "Flexlevel nand flash storage system design to reduce ldpc latency," *IEEE Transactions on Computer-Aided Design of Integrated Circuits and Systems*, vol. 36, no. 7, pp. 1167–1180, July 2017.

[143] L. Zuolo, C. Zambelli, A. Marelli, R. Micheloni, and P. Olivo, "Ldpc soft decoding with improved performance in 1x-2x mlc and tlc nand flash-based

solid state drives," *IEEE Transactions on Emerging Topics in Computing*, vol. PP, no. 99, pp. 1–1, 2017.

[144] J. Spinner and J. Freudenberger, "A decoder with soft decoding capability for high-rate generalized concatenated codes with applications in non-volatile flash memories," in *Proceedings of 30th Symposium on Integrated Circuits and Systems Design (SBCCI), Fortaleza, Brazil*, Sept. 2017.

[145] D Chase, "Class of algorithms for decoding block codes with channel measurement information," *IEEE Transactions on Information Theory*, pp. 170–182, 1972.

[146] T. Kaneko, T. Nishijima, H. Inazumi, and S. Hirasawa, "An efficient maximum-likelihood-decoding algorithm for linear block codes with algebraic decoder," *IEEE Transactions on Information Theory*, vol. 40, no. 2, pp. 320–327, Mar 1994.

[147] T. Kaneko, T. Nishijima, and S. Hirasawa, "An improvement of soft-decision maximum-likelihood decoding algorithm using hard-decision bounded-distance decoding," *IEEE Transactions on Information Theory*, vol. 43, no. 4, pp. 1314–1319, Jul 1997.

[148] H. Yamamoto and K. Itoh, "Viterbi decoding algorithm for convolutional codes with repeat request," *IEEE Transactions on Information Theory*, vol. 26, no. 5, pp. 540–547, Sep 1980.

[149] G. Forney, "Exponential error bounds for erasure, list, and decision feedback schemes," *IEEE Transactions on Information Theory*, vol. 14, no. 2, pp. 206–220, Mar 1968.

[150] Stelios Korkotsides, Georgios Bikas, Efstratios Eftaxiadis, and Theodore Antonakopoulos, "Ber analysis of mlc nand flash memories based on an asymmetric pam model," in *Communications, Control and Signal Processing (ISCCSP), 2014 6th International Symposium on*. IEEE, 2014, pp. 558–561.

[151] Yu Cai, Onur Mutlu, Erich F Haratsch, and Ken Mai, "Program interference in mlc nand flash memory: Characterization, modeling, and mitigation," in *Computer Design (ICCD), 2013 IEEE 31st International Conference on*. IEEE, 2013, pp. 123–130.

[152] H. Yassine, J. Coon, M. Ismail, and H. Fletcher, "Towards an analytical model of NAND flash memory and the impact on channel decoding," in *IEEE International Conference on Communications (ICC)*, May 2016, pp. 1–6.

[153] D. h. Lee and W. Sung, "Estimation of NAND flash memory threshold voltage distribution for optimum soft-decision error correction," *IEEE Transactions on Signal Processing*, vol. 61, no. 2, pp. 440–449, Jan 2013.

[154] N Tendolkar and C Hartmann, "Generalization of Chase algorithms for soft decision decoding of binary linear codes," *IEEE transactions on information theory*, vol. 30, no. 5, pp. 714–721, 1984.

[155] C. M. Bishop, *Pattern Recognition and Machine Learning*, (Information Science and Statistics), Springer, New York, 2006.

Printed in the United States
By Bookmasters